SPECIAL REP(

DESIGNING SAFER ROADS

Practices for
Resurfacing,
Restoration, and
Rehabilitation

Transportation Research Board
National Research Council
Washington, D.C. 1987

Transportation Research Board Special Report 214

mode
1 highway transportation

subject areas
21 facilities design
51 transportation safety
52 human factors

Transportation Research Board publications are available by ordering directly from TRB. They may also be obtained on a regular basis through organizational or individual affiliation with TRB; affiliates or library subscribers are eligible for substantial discounts. For further information, write to the Transportation Research Board, National Research Council, 2101 Constitution Avenue, N.W., Washington, D.C. 20418.

Printed in the United States of America

NOTICE
The project that is the subject of this publication was approved by the Governing Board of the National Academy of Sciences, the National Academy of Engineering, and the Institute of Medicine. The members of the committee responsible for the publication were chosen for their special competence and with regard for appropriate balance.

This report has been reviewed by a group other than the authors according to procedures approved by a Report Review Committee consisting of members of the National Academy of Sciences, the National Academy of Engineering, and the Institute of Medicine.

This study was sponsored by the Federal Highway Administration of the U.S. Department of Transportation.

Library of Congress Cataloging-in-Publication Data

National Research Council (U.S.). Transportation Research Board.
  Designing safer roads : practices for resurfacing, restoration, and rehabilitation.
  p. cm. — (Special report / Transportation Research Board, National Research Council ; 214)
  Includes bibliographies.
  ISBN 0-309-04453-7
  1. Roads—Maintenance and repair. 2. Roads—Design. I. Title.
II. Series: Special report [National Research Council (U.S).
Transportation Research Board] ; 214.
TE220.N37 1987
625.7'6—dc19   87-15337
                                                             CIP

COMMITTEE FOR THE STUDY OF GEOMETRIC DESIGN STANDARDS FOR HIGHWAY IMPROVEMENTS

PETER G. KOLTNOW, American Trucking Associations, Alexandria, Virginia *Co-Chairman*
HERBERT H. RICHARDSON, The Texas A&M University System, College Station, *Co-Chairman*
ROY W. ANDERSON, TranSafety, Inc., Springfield, Virginia
LEONARD EVANS, General Motors Research Laboratories, Warren, Michigan
JOHN C. GLENNON, John C. Glennon Chartered, Prairie Village, Kansas
EZRA HAUER, University of Toronto, Ontario
W. RONALD HUDSON, University of Texas, Austin
JACK T. KASSEL, Sacramento, California
JAMES L. MARTIN, Fresno, California
BROOKS O. NICHOLS, Arkansas State Highway and Transportation Department, Little Rock
BRIAN O'NEILL, The Insurance Institute for Highway Safety, Washington, D.C.
ROBERT H. RAYMOND, JR., Pennsylvania Department of Transportation, Harrisburg
JOHN H. SHAFER, New York State Department of Transportation, Albany
RICHARD R. STANDER, JR., Mansfield Asphalt Paving Company, Mansfield, Ohio
JAMES I. TAYLOR, University of Notre Dame, Indiana
E. DEAN TISDALE, Idaho Transportation Department, Boise

Liaison Representatives

DAVID J. HENSING, American Association of State Highway and Transportation Officials, Washington, D.C.
MICHELE A. MCMURTRY, National Transportation Safety Board
JEAN SCHRAG-LAUVER, Senate Environment and Public Works Committee
SEPPO I. SILLAN, Federal Highway Administration, U.S. Department of Transportation
RICHARD V. TEARLE, House Committee on Public Works and Transportation
JENNIFER WISHART, Congressional Budget Office
DAVID K. WITHEFORD, Transportation Research Board
CLYDE E. WOODLE, JR., House Committee on Public Works and Transportation

Transportation Research Board Staff

ROBERT E. SKINNER, Director for Special Projects
HARRY S. COHEN, Senior Program Officer
JOSEPH R. MORRIS, Senior Program Officer
JOHN A. DEACON, Consultant
RICHARD MARGIOTTA, Research Associate
MALCOLM QUINT, Research Associate
EDYTHE TRAYLOR CRUMP, Senior Editor

Preface

In response to a provision in the Surface Transportation Assistance Act of 1982, the Secretary of Transportation, acting through the Federal Highway Administration, requested the National Academy of Sciences to study the safety cost-effectiveness of highway geometric design standards and recommend minimum standards for resurfacing, restoration, and rehabilitation (RRR) projects on existing federal-aid highways, except freeways. Specifically, the act called for the Secretary of Transportation to enter into arrangements with the National Academy of Sciences to

> conduct a study of the safety cost-effectiveness of geometric design criteria of standards currently in effect for construction and reconstruction of highways, other than highways access to which is fully controlled, to determine the most appropriate minimum standards to apply to resurfacing, restoration, and rehabilitation projects on such highways . . . and to propose standards to preserve and extend the service life of such highways and enhance highway safety.

To carry out the study, the National Research Council, the principal operating agency of the National Academy of Sciences and the National Academy of Engineering, assembled a committee of 16 experts in the various disciplines needed to develop and apply geometric design standards and assess their impact on safety, highway serviceability, cost, environment, and system administration. Committee members included individuals with experience in highway design, traffic engineering, highway safety, accident analysis, highway construction, statistics, economics, highway administration, and law.

The committee began its work with a review of RRR practices in state and local highway agencies. Committee staff visited the state highway agency and the Federal Highway Administration offices in each of the 15 states selected for case studies and conducted telephone interviews with local highway officials representing 16 counties, 20 cities, and 3 metropolitan planning organizations. Federal, state, and local officials provided valuable information on the types of projects funded with federal aid, procedures used to select RRR projects, current design standards and their use, and the ways in which safety needs are taken into account.

The study committee sponsored critical reviews of prior research on the safety effects of key highway features and special research projects on pavement edge drops and roadside safety. The critical reviews and findings from the special research projects were used to make judgments about relationships between safety and key highway features. For several design features, the committee found sufficient evidence to support quantitative relationships between safety and design improvements. However, these relationships must be viewed as approximate in nature. Although the relationships are based on the best available data, they could be substantially changed by the results of future research.

In addition, the study committee developed relationships between cost and key highway features. These relationships are based on an examination of published cost data, cost records, and cost-estimating procedures for a sample of highway agencies throughout the country. The cost relationships provide estimates of typical costs for making geometric design improvements on RRR projects. However, the cost for a given improvement can vary considerably from site to site because of variations in site conditions, labor and material costs, design practices, and project scale. Thus, actual costs could be much greater or less than estimates developed using the cost relationships.

The safety and cost relationships were used to assess the safety cost-effectiveness of geometric design standards. The added cost per accident eliminated that can be expected for improvements to highway geometry was estimated for illustrative projects. When system data were available for existing highway conditions, the study committee examined the effects of alternative RRR standards on systemwide safety and the total expenditure needed to meet the standard on a nationwide basis or for selected states.

Drawing primarily on case studies of current RRR practices and analyses of safety cost-effectiveness, the committee has recommended a variety of practices that encompass the entire RRR process but with special focus on design. In selected instances, federal, state, and local highway agencies can use the recommendations, along with published manuals, design aids, and local experience to develop or modify minimum design standards for RRR projects. For federal-aid RRR work, the Secretary of Transportation is required by statute to ensure that projects are designed and constructed in accordance with standards that extend the service life of highways and enhance highway safety. To accomplish this, the Secretary, acting through the Federal Highway Administration, must either set nationwide RRR standards or approve standards adopted by individual states. In either case, the committee's recommendations provide guidance. In addition, the committee has recommended various research and training activities that federal and state highway agencies can use to improve their ability to enhance safety through RRR projects.

The study was performed under the overall supervision of Dr. Damian J. Kulash and Robert E. Skinner, Jr., the former and current Directors for Special Projects. Robert E. Skinner, Jr., directed the project staff. Dr. Harry Cohen, Joseph R. Morris, Dr. John A. Deacon, Richard Margiotta, and Malcolm Quint made significant contributions.

Special appreciation is expressed to Nancy A. Ackerman, TRB Publications Manager, and Edythe T. Crump, Senior Editor, for editing the final report and to Marguerite E. Schneider, Frances E. Holland, and Margaret M. Sheriff for typing the many drafts and the final manuscript.

Contents

EXECUTIVE SUMMARY 1

1 GEOMETRIC DESIGN STANDARDS FOR RESURFACING, RESTORATION, AND REHABILITATION PROJECTS: BACKGROUND AND ISSUES 14
    Introduction, 14
    Evolution of Federal Highway Policy, 16
    Federal-Aid Highway Program, 18
    Geometric Design Standards and Federal Rulemaking, 24
    Key Issues, 30
    References, 32

2 STATE AND LOCAL PROCEDURES FOR SELECTION, DESIGN, AND CONSTRUCTION OF HIGHWAY IMPROVEMENT PROJECTS 35
    Review of RRR Practices: Information Sources, 37
    State RRR Programs, 38
    Local RRR Programs, 66
    Summary of Findings, 72
    References, 75

3 RELATIONSHIPS BETWEEN SAFETY AND GEOMETRIC DESIGN 76
    Application of Safety Relationships to Design Standards, 76
    Relationships Between Safety and Key Road Features, 78
    Low-Cost Safety Measures, 100
    Effect of Changing Vehicle Fleet, 102
    Roadway Consistency, 104
    Summary, 105
    References, 106

4   **RELATIONSHIPS BETWEEN HIGHWAY COSTS AND GEOMETRIC DESIGN** 110
    Cost Relationships—Problems and Limitations, 110
    Typical RRR Project Costs, 113
    Added Project Costs for Geometric Improvements, 116
    Right-of-Way Requirements, 125
    Maintenance Cost Implications, 126
    Summary, 129
    References, 129

5   **SAFETY COST-EFFECTIVENESS OF GEOMETRIC DESIGN STANDARDS** ... 131
    Earlier Studies of Safety Cost-Effectiveness in Highway Design, 132
    Scope and Framework of Cost-Effectiveness Analyses, 133
    Safety-Cost Trade-Offs, 136
    Safety-Preservation Trade-Offs, 166
    Summary of Findings, 170
    References, 172

6   **TORT LIABILITY AND GEOMETRIC DESIGN** 174
    Background on Tort Liability, 175
    Implications for RRR Design Standards and Practices, 178
    Summary, 183
    References, 184

7   **FINDINGS AND RECOMMENDED DESIGN PRACTICES FOR RESURFACING, RESTORATION, AND REHABILITATION PROJECTS** 186
    Findings, 186
    Safety-Conscious Design Process, 190
    Design Practices for Key Highway Features, 193
    Other Design Procedures and Assumptions, 204
    Planning and Programming RRR Projects, 207
    Safety Research and Training, 208
    References, 212

APPENDIX A   Summary Comparison of Nonfreeway Geometric Design Standards and Guidelines 213

APPENDIX B   Case Study State and Local RRR Programs 220

APPENDIX C   Summary of Detailed Safety Relationships 248

APPENDIX D   Relationship Between Accidents and Horizontal Curvature 256

APPENDIX E   Relationship Between Accidents and Sight Distance at Crest Vertical Curves 265

APPENDIX F	Relationship Between Accidents and Specific Roadside Features 270
APPENDIX G	Physical and Operational Features Affecting Safety at Intersections 286
APPENDIX H	Highway Accidents on the Federal-Aid System 292
APPENDIX I	Initial Cost to Flatten Highway Curves 296
APPENDIX J	Relationship Between Cost per Accident Eliminated and Benefit-Cost Ratio Approaches 301
APPENDIX K	Effects of Lane and Shoulder Widths on Travel Time 303
APPENDIX L	Alternative Lane and Shoulder Width Standards Used in System-Level Analyses 308

STUDY COMMITTEE BIOGRAPHICAL INFORMATION 312

Executive Summary

In response to a provision of the Surface Transportation Assistance Act of 1982, the Secretary of Transportation, acting through the Federal Highway Administration, requested the National Academy of Sciences to study the safety cost-effectiveness of geometric design standards and recommend minimum standards for resurfacing, restoration, and rehabilitation (RRR) projects on existing federal-aid highways, except freeways. RRR projects may include resurfacing and other pavement repairs, minor widening of lanes and shoulders, minor alterations to vertical and horizontal alignment, bridge improvements, and removal of roadside hazards.

Until 1976, federal highway funds could be used only for the construction of new highways or complete reconstruction of existing highways. This policy was changed by the Federal-Aid Highway Act of 1976, which authorized state and local highway agencies to use federal aid for RRR projects on existing federal-aid highways.

RRR projects can extend the service life of existing highways through pavement and other repairs and at the same time improve highway safety by making selective improvements to highway geometry and other roadside features. Striking a balance between preservation and safety improvements on RRR projects has proved controversial, however.

The controversy has centered on which minimum geometric design standards should be applied to RRR projects to qualify for federal aid. Some highway organizations have contended that pavement repairs alone enhance safety and that additional safety improvements would greatly increase project costs and delay improvements to many miles of deteriorating highways.

Safety organizations, on the other hand, have viewed the federal RRR program as an opportunity to make long-needed safety improvements to older highways at the same time as pavement repairs are made. These organizations have viewed the flexible RRR standards proposed by some highway agencies as too lenient and have favored a more rigorous, safety-oriented design process.

A study committee of 16 individuals with expertise in highway safety, design, and administration conducted case studies of current RRR design practices, reviewed current knowledge about relationships between geometric design and safety, and analyzed the cost and safety trade-offs of geometric improvements to existing highways. These activities led to a number of findings concerning the effects of RRR projects on highway safety. On the basis of these findings, the study committee has recommended a variety of practices that will increase the safety cost-effectiveness of RRR projects.

FINDINGS

Resurfacing, restoration, and rehabilitation projects enable highway agencies to improve highway safety by selectively upgrading existing highway and roadside features without the cost of full reconstruction. For example, widening lanes and shoulders on two-lane rural highways on the federal-aid systems alone could save approximately 1,000 lives and prevent nearly 30,000 injuries each year.

Federal-aid RRR projects usually enhance safety. Moreover, since 1982 when Congress declared that RRR project objectives include both the extension of highway service life and the enhancement of safety, highway agencies generally have paid increasing attention to safety. Nevertheless, many opportunities for low-cost safety improvements are neglected, and RRR funds currently spent for safety improvements could be redirected for greater systemwide safety gains. A number of factors are responsible for this situation:

- *RRR design practices vary widely from agency to agency.* Some highway agencies follow exemplary practices to address safety needs; others do not place enough emphasis on safety.
- *RRR projects are initiated primarily to address pavement repair and rehabilitation needs.* Safety needs are often not addressed until little time remains to accommodate geometric improvements that require additional time for design or right-of-way acquisition.
- *Federal-aid RRR projects frequently widen lanes and shoulders but seldom reconstruct sharp curves or replace bridges with narrow decks.* Because there is a higher concentration of accidents at curves and bridges,

improvements at these locations can sometimes be more safety cost-effective than routine cross-section improvements.

• *Not enough is known about the safety gains that will occur after the geometry of existing highways is improved or other safety-oriented improvements are made.* Available information is not always in the hands of designers, or in a form that can be applied without ambiguity. Also, past studies of the safety effects of geometric design improvements frequently lacked rigorous statistical controls, a shortcoming that severely limits the accuracy of study results.

• *Engineers who administer state traffic and safety programs seldom participate in the design of RRR projects.* They are usually the agency staff members most knowledgeable about accident data and special safety measures.

Design standards alone cannot address these factors that collectively limit the safety gains of federally funded RRR projects. Within the overall process of planning, selecting, and designing RRR projects, the influence of safety standards is small. RRR standards, which can affect only a few key design features, cannot be tailored to fit all possible, or even most, circumstances encountered in a given state or at a specific site.

Consequently, a variety of practices are recommended that encompass the RRR process but with special focus on design. In selected instances, federal, state, and local highway agencies can use the recommendations, along with published manuals, design aids, and local experience to develop or modify minimum design standards for RRR projects. For federal-aid RRR work, the Secretary of Transportation is required by statute to ensure that projects are designed and constructed in accordance with standards that extend the service life of highways and enhance highway safety. To accomplish this, the Secretary, acting through the Federal Highway Administration, must either set nationwide RRR standards or approve standards adopted by individual states. In either case, the committee's recommendations provide guidance.

The recommended practices also provide guidance on the planning and programming of RRR projects, existing conditions that warrant special design analyses, safety improvements that should routinely be considered on RRR projects, and training and research. These practices are intended to develop more safety-conscious design. This will enhance highway safety nationwide by taking advantage of low-cost opportunities to improve safety and selecting the most safety cost-effective improvements.

If these recommendations are followed for federal-aid projects on nonfreeway highways, project spending for lane and shoulder widening will generally decline and spending for alignment, bridge, roadside, and intersection improvements, as well as project design, should increase. In some states these shifts may decrease RRR project costs; in others they will increase costs. Nationwide, the typical project cost will probably increase slightly but not enough to measurably affect RRR pavement repair and preservation activities.

RECOMMENDATIONS

Study recommendations are organized into five categories (Table ES-1):

1. Safety-conscious design process,
2. Design practices for key highway features,
3. Other design procedures and assumptions,
4. Planning and programming RRR projects, and
5. Safety research and training.

TABLE ES-1 Organization of Study Recommendations

Safety-Conscious Design Process
   1. Assessment of Site Conditions Affecting Safety
   2. Determination of Project Scope
   3. Documentation of the Design Process
   4. Review by Traffic and Safety Engineers
Design Practices for Key Highway Features
   5. Minimum Lane and Shoulder Widths
 6, 7. Horizontal Curvature and Superelevation
   8. Vertical Curvature and Stopping Sight Distance
   9. Bridge Width
  10. Sideslopes and Clear Zones
  11. Pavement Edge Drop and Shoulder Type
  12. Intersections
  13. Normal Pavement Crown
Other Design Procedures and Assumptions
  14. Traffic Volume Estimates for Evaluating Geometric Improvements
  15. Speed Estimates for Evaluating Geometric Improvements
  16. Design Values for Geometric Improvements
  17. Design Exceptions
Planning and Programming RRR Projects
  18. Screening of Highways Programmed for RRR Projects
  19. Assessment of the Systemwide Potential for Improving Safety
Safety Research and Training
  20. Special Task Force to Assess Highway Safety Needs and Priorities
  21. Compendium of Information on Safety Effects of Design Improvements
  22. Increased Research on the Relationships Between Safety and Design
  23. Safety Training Activities for Design Engineers

These recommendations apply to nonfreeway RRR projects whether or not they are funded with federal aid. For federal-aid RRR projects in particular, the Secretary of Transportation, through the Federal Highway Administration, should take the necessary steps to implement the recommendations in the first three categories—safety-conscious design process, design practices for key highway features, and other design procedures and assumptions. Taken together, these recommendations comprise a practical national policy on RRR project design that will be more safety-cost effective and comprehensive than an extensive set of rigid minimum standards.

The fourth category, planning and programming RRR projects, is directed to state and local highway agencies that have the authority to perform these functions for federal-aid projects without federal oversight. The final category, safety research and training, is directed to the larger highway community with specific recommendations intended for the Congress, the Federal Highway Administration (FHWA), the American Association of State Highway and Transportation Officials (AASHTO), and state and local highway agencies.

Safety-Conscious Design Process

Significant improvements in safety are not automatic by-products of RRR projects; safety must be systematically designed into each project. Highway designers must deliberately seek opportunities specific to each project and apply sound safety and traffic engineering principles. Designers of RRR projects work with existing highways whose design and operational characteristics can be observed and measured; yet not all highway agencies take advantage of these favorable circumstances. Greater attention to safety, along with greater documentation of the design process improves design decisions. Highway agencies should review and revise their design practices to incorporate the following steps.

• *Assess current conditions:* At the beginning of RRR project design, highway designers should assess existing physical and operational conditions affecting safety by using accident data, site inspections, and measurement of existing design and traffic characteristics.

• *Determine project scope:* In addition to pavement repairs and geometric improvements, designers of RRR projects should consider and, where appropriate, incorporate other intersection, roadside, and traffic control improvements that may enhance safety.

• *Document the design process:* Before developing construction plans and specifications, designers should prepare a safety and design report that covers existing design and operational characteristics, accident history, applicable design standards, identified safety problems and related design options, rationale for any proposed design exceptions, and the recommended design.

• *Review the design:* Traffic and safety engineers should review safety and design reports, as well as proposed RRR designs, before final approval.

Although many state highway agencies already incorporate one or more of these steps in their design process, most will have to modify their process to include them all.

Design Practices for Key Highway Features

Designers use minimum RRR geometric design standards to determine whether a particular geometric feature must be upgraded as part of a RRR project. Numerical minimum RRR standards are warranted for nationwide use when the following conditions are met:

- Trade-offs between safety and performance against cost can be evaluated quantitatively, and conclusions can be drawn about the safety cost-effectiveness of different standards generally applicable regardless of the project.
- Standards would help refocus RRR expenditures on more safety cost-effective geometric improvements.
- Standards would simplify parts of the design process and FHWA approval procedures, freeing design resources for the analysis of site improvements not covered by numerical standards.

Lane and shoulder widths on two-lane rural highways meet these conditions, and minimum values are recommended. Two-lane rural highways account for about three-fourths of all nonfreeway, federal-aid highway mileage, and lane and shoulder widths are particularly important because they can affect highway safety and cost over the length of the highway.

When these conditions are not met, for other key features or categories of highways, other design practices are recommended that will help achieve the same safety objectives as minimum standards. These recommended practices specify threshold conditions that warrant detailed evaluation of particular improvements, improvements that should routinely be made or evaluated on RRR projects, or design policies that should be developed on a state-by-state basis.

Minimum Lane and Shoulder Widths

Minimum lane and shoulder width values are recommended that FHWA and state highway agencies can use to set minimum RRR design standards (Table ES-2). These recommended values are similar to the minimum lane and shoulder width values proposed by the FHWA in 1978 but include several modifications to improve safety cost-effectiveness. Most important, the aver-

TABLE ES-2 Recommended Minimum Lane and Shoulder Width Values for Two-Lane Rural Highways

Design Year Volume (ADT)	Running Speed[a] (mph)	10 Percent or More Trucks[b]		Less Than 10 Percent Trucks	
		Lane Width	Combined Lane and Shoulder Width[c]	Lane Width	Combined Lane and Shoulder Width[c]
1–750	Under 50	10	12	9	11
	50 and over	10	12	10	12
751–2,000	Under 50	11	13	10	12
	50 and over	12	15	11	14
Over 2,000	All	12	18	11	17

[a]Highway segments should be classified as "under 50" only if most vehicles have an average speed of less than 50 mph over the length of the segment.
[b]For this comparison, trucks are defined as heavy vehicles with six or more tires.
[c]One foot less for highways on mountainous terrain.

age daily traffic (ADT) ranges are adjusted so that a larger number of roads with high ADT and fewer roads with low ADT would be improved. Lane and shoulder width improvements are more cost-effective on high-volume roads than on low-volume roads.

In terms of cost per accident eliminated, the recommended values are more cost-effective than other standards proposed for nationwide use. For all federal-aid, two-lane rural highways combined, the recommended minimum values imply approximately the same overall investment as the FHWA standards proposed in 1978—a total of roughly $13 billion if all of the lane and shoulder improvements were made at current cost levels. Application of these values, however, would eliminate about 10,000 (40 percent) additional accidents annually.

Less is known about the safety cost-effectiveness of widening urban and multilane rural highways, and minimum values that highway agencies can adopt as standards have not been proposed.

Horizontal Curvature and Superelevation

Current RRR standards and practices generally emphasize lane and shoulder width improvements and do not pay enough attention to alignment improvements. At traffic volumes greater than 750 vehicles per day, reconstruction of horizontal curves can be more safety cost-effective than lane and shoulder widening and can reduce vehicle operating costs and travel time.

Because of the variability in project costs (and safety cost-effectiveness) for reconstructing similar curves, minimum geometric standards are inappropriate. Nevertheless, highway agencies should evaluate the safety benefits and added costs of curve reconstruction when there is a reasonable possibility that reconstruction will be safety cost-effective.

The study recommends that highway agencies

- Evaluate the reconstruction of horizontal curves when the design speed of the existing curve is more than 15 mph below the running speeds of approaching vehicles and the average daily traffic volume is greater than 750 vehicles per day.
- Increase the superelevation of horizontal curves whenever the design speed of an existing curve is below the running speeds of approaching vehicles and the existing superelevation is below the allowable maximum specified by AASHTO new construction policies.

In many cases, safety can be improved at horizontal curves without costly reconstruction. Where reconstruction is unwarranted, highway agencies should evaluate less costly safety measures such as widening lanes, widening or paving shoulders, flattening steep sideslopes, removing or relocating roadside obstacles, and installing traffic control devices.

Vertical Curvature and Stopping Sight Distance

Reconstruction of vertical curves at hill crests to increase stopping sight distance may be safety cost-effective at average daily traffic volumes greater than 1,500 vehicles per day depending on site conditions. Generally, to be safety cost-effective, vertical curve reconstruction must correct a substantial sight distance restriction that affects drivers' ability to anticipate a hazardous situation—turning vehicles, sharp curves, or other conditions that demand specific driver responses.

Highway agencies should evaluate the reconstruction of hill crests when *(a)* the hill crest hides from view major hazards such as intersections, sharp horizontal curves, or narrow bridges; *(b)* the average daily traffic is greater than 1,500 vehicles per day; and *(c)* the design speed of the hill crest (based on the minimum stopping sight distance provided) is more than 20 mph below the running speeds of vehicles on the crest.

Whether or not an evaluation of hill crest reconstruction is required, designers should examine the nature of potential hazards hidden by a hill crest and consider other options such as removing the hazards or providing warning signs.

Bridge Width

The safety cost-effectiveness of bridge width improvements depends on the usable width of the bridge, the width of approach lanes, traffic volumes, and the length of the bridge. Highway agencies should evaluate bridge replacement or widening in situations in which bridge width improvements might be justified on the basis of safety cost-effectiveness—bridges less than 100 ft long with usable widths less than the values given in Table ES-3. At low traffic volumes the recommended values are similar to those proposed by the 1978 FHWA standards, and at high traffic volumes they are similar to those specified by the AASHTO policy for bridges to remain in place on arterial highways.

TABLE ES-3 Usable Bridge Widths Below Which Bridge Replacement or Widening Should Be Evaluated (If Bridge is Less Than 100 ft Long)

Design Year Volume (ADT)	Usable Bridge Width (ft)[a]
0–750	Width of approach lanes
751–2,000	Width of approach lanes plus 2 ft
2,001–4,000	Width of approach lanes plus 4 ft
Over 4,000	Width of approach lanes plus 6 ft

[a] If lane widening is planned as part of the RRR project, the usable bridge width should be compared with the planned width of the approaches after they are widened.

Whether or not evaluation of bridge widening is warranted, designers should consider installing transition guardrails at bridge approaches, rehabilitated or new bridge rails, and warning signs.

Sideslopes and Clear Zones

Roadside characteristics are important in determining the overall level of safety provided by a highway. Accident rates are lower and accidents are less severe on highways with gentle sideslopes and few obstacles near the roadway. Despite these findings, the study revealed no basis for nationwide standards addressing either sideslopes or clear zone width. The safety cost-effectiveness of particular roadside improvements appears highly dependent on site-specific conditions and interactions between different roadside features.

The study recommends that highway agencies develop and apply their own procedures for identifying and selecting sideslope and clear zone width improvements on RRR projects. These procedures should encourage the following:

- Flatten sideslopes of 3:1 or steeper at locations where run-off-road accidents are likely to occur (e.g., on the outside of sharp horizontal curves);
- Retain current slope widths (without steepening sideslopes) when widening lanes and shoulders unless warranted by special circumstances; and
- Remove, relocate, or shield isolated roadside obstacles.

Pavement Edge Drop and Shoulder Type

Pavement edge drops (i.e., vertical drops or ruts) often develop between the pavement surface and adjacent unpaved shoulders or roadsides. These drops can prevent drivers whose vehicles cross over the lane edge from successfully returning to their original lane without encroaching on an opposing lane or losing control.

Research sponsored as part of this study indicated that pavement edge drop hazards are greater than previously believed. However, no basis exists for estimating how often pavement edge drops contribute to accidents or the cost and safety trade-offs involved in preventing or correcting them.

Depending on the type of shoulder construction, resurfacing can increase the likelihood that edge drops will develop later and require repeated maintenance to correct. To reduce pavement edge drop hazards on highways with narrow unpaved shoulders, highway agencies should either

- Selectively pave shoulders at points where out-of-lane vehicle excursions and pavement edge drop problems are likely to develop (e.g., at horizontal curves); or
- Construct a beveled or tapered pavement edge shape at these points.

Intersections

Reliable information about the cost and safety trade-offs of individual intersection improvements is generally unavailable because of the large number of physical and operational features affecting intersection safety and because intersection projects typically address multiple intersection safety problems simultaneously. Nevertheless, many intersection improvements can be made at relatively low cost and are safety cost-effective, particularly at higher traffic volumes.

Nationwide numerical standards are inappropriate for RRR projects. Therefore, designers must tailor intersection improvements to site-specific conditions and rely heavily on professional judgment and experience. To facilitate this, state highway agencies should develop criteria for identifying intersections that warrant careful evaluation and checklists of improvements to be considered.

Other Design Procedures and Assumptions

Different highway agencies may design RRR projects differently even when their minimum RRR standards are the same and project conditions are practically identical. Four procedures are recommended that will encourage a more uniform application of RRR standards and a more consistent approach to safety.

- *Design traffic volume:* The design traffic volume for a given highway feature should match the average traffic anticipated over the expected performance period of that feature. Most state highway agencies use current-year traffic even though the expected performance period of the pavement rehabilitation work is 5 to 15 years and the performance period for geometric improvements may exceed 25 years.
- *Speed:* When determining whether geometric improvements are warranted for features for which vehicle speed is a factor, highway agencies should estimate actual running speeds in a manner appropriate for the feature under consideration. For example, for horizontal curves, designers should use the 85th percentile speed of vehicles approaching the curve, estimated at a point where drivers have not yet reduced speed.
- *Design values:* When selecting design values for geometric improvements, highway agencies should estimate the incremental safety cost-effectiveness of improvements that exceed the minimum standard and should consider overall highway geometry, design of adjacent sections, and expected trends in traffic growth and truck use. Improvements beyond the RRR minimum standards may not be cost-effective and may create inconsistencies between the level of safety provided by the features improved on the RRR project and the features unimproved.
- *Design exceptions:* Highway agency requests for an exception to a design standard should explicitly address the expected safety consequences, along with cost and other impacts. The cited justifications for exceptions are often imprecise and vary from state to state.

Planning and Programming RRR Projects

Highway agencies select RRR projects primarily on the basis of pavement repair needs and seldom consider safety needs until preliminary design begins.

As a result, most highway agencies have not determined where geometric improvements to existing highways would have the greatest safety payoffs and be the most safety cost-effective. Moreover, when additional right-of-way is needed for a RRR geometric improvement, highway agencies must often delay the project or neglect the improvement.

Given current budget levels and existing highway conditions, pavement repair needs will continue to be the dominant factor in the selection and scheduling of RRR projects. Nevertheless, highway agencies should begin to take safety into account earlier in the overall RRR process.

- *Systemwide safety planning:* Highway agencies should periodically assess the systemwide potential for improving safety through upgraded design to help guide project programming and design practices.
- *Expedite right-of-way acquisitions:* Highway agencies should review highways programmed for RRR projects to identify locations where design and right-of-way acquisition should be expedited.

Safety Research and Training

Despite more than one-half century of modern road building, knowledge of the safety consequences of highway design decisions is limited. Except for a modest FHWA research program and occasional research studies sponsored by the National Cooperative Highway Research Program, few opportunities exist for coordinated, purposeful safety research aimed at improving this knowledge. To a large extent, the highway community has relied on uncoordinated research without rigorous statistical controls to expand knowledge about the safety effects of highway design. Conflicting research findings are often left unresolved.

Equally serious, current knowledge about the safety effects of highway design is inconsistently applied. Although designers can rely on standards and design aids in many instances, some decisions must be based on site-specific circumstances and judgment. Designers often lack the capability or time to apply the existing safety knowledge.

Additional resources must be devoted to safety research, training, and design to expand knowledge about the relationships between safety and highway design and take steps to assure that this knowledge is properly applied. The study committee believes that the payoff in long-term highway safety gains will be worth the added cost.

The recommendations that follow offer the first step toward meeting long-term research and training needs and also suggest steps that can be taken now to improve research and training.

- *Special safety task force:* Congress should direct the Secretary of Transportation to establish a special task force to assess highway safety engineering needs and to establish research, education, and funding priorities to meet these needs.
- *Safety compendium:* The Federal Highway Administration should develop, distribute, and periodically update a compendium that reports the most probable safety effects of improvements to key highway design features and identifies the principal gaps in current knowledge.
- *Increased research:* The Federal Highway Administration and the National Cooperative Highway Research Program should increase research on the relationships between safety and highway design.
- *Training:* The Federal Highway Administration, the American Association of State Highway and Transportation Officials, state and local highway agencies, and other organizations of public works professionals should support continuing training activities to keep design engineers abreast of safety-conscious design.

1
Geometric Design Standards for Resurfacing, Restoration, and Rehabilitation Projects: Background and Issues

INTRODUCTION

Since 1976 when the U.S. Congress first authorized federal aid for resurfacing, restoration, and rehabilitation (RRR) work, questions about geometric design standards have persisted:

• What improvements to existing highways yield the greatest safety gains in relation to cost?
• How can the overall process of selecting, designing, and constructing RRR projects take advantage of these opportunities for safety improvements?
• How much federal aid should be used for resurfacing and other pavement repairs that preserve and extend the service life of existing highways?

Many state highway organizations have viewed the federal RRR program primarily as a means of addressing critical pavement repair needs. During the 1970s these needs mounted as construction costs escalated and state highway revenues leveled off or declined. As a result, flexible geometric design standards were preferred for RRR projects because stringent standards such as those required for new construction or full reconstruction would, if followed rigorously, dramatically increase project costs. State highway organizations believed that such increases would inevitably lead to the concentration of available funds on a small number of projects, leaving unattended many federal-aid highways in need of pavement repair and meeting neither preservation nor safety objectives.

Safety organizations, on the other hand, have viewed the federal RRR program as an opportunity to make long-needed safety improvements to older highways at the same time as pavement repairs are made. These organizations have viewed the flexible RRR standards proposed by some highway agencies as too lenient and have favored a more rigorous, safety-oriented design process.

RRR projects may include resurfacing, pavement structural and joint repairs, widening of lanes and shoulders, selected alterations to vertical and horizontal alignment, bridge repairs, and removal of roadside hazards. The federal government considers more extensive improvements reconstruction; lesser repairs are viewed as maintenance and therefore ineligible for federal aid.

Nearly $5 billion, about 10 percent of highway expenditures by all levels of government, are spent annually on non-Interstate RRR repairs, a share that will likely increase as highway programs continue to shift emphasis from construction to preservation. Federal-aid highways account for approximately one-half of the $5 billion spent. From a safety standpoint, about 30,000 persons are killed each year on non-Interstate federal-aid highways, which amounts to two-thirds of all U.S. traffic fatalities, and almost 2 million persons are injured each year.

Until 1976, state and local governments were responsible for undertaking and financing RRR work, and they made their own judgments about the priorities of pavement repair and geometric improvements. Reacting to widespread concern over the deteriorating condition of the nation's highway system in 1976, Congress authorized the use of federal-aid construction funds for RRR projects. Because no other standards existed, the Federal Highway Administration (FHWA), which oversees the federal highway program, initially applied new construction standards, expecting to later adopt special RRR standards for older nonfreeway highways. Design exceptions were permitted on a case-by-case basis to accommodate difficult situations such as widening roads in urban areas or straightening roads in mountainous regions. Because many of the roads involved were built many years ago to different standards, geometric features such as shoulder widths or curve radii are not uniform. Imposing new construction standards on a nationwide basis resulted in a large number of design exceptions in cases in which upgrading to these standards would have been extraordinarily expensive.

Because of the divergent views on standards and the controversy that arose, FHWA never adopted a set of special RRR standards for nationwide use. Instead it adopted a flexible approach that permits states to develop and apply their own RRR standards subject to federal approval, or to continue to use new construction standards. This action failed to silence the debate over RRR standards as evidenced by the congressional mandate for this study, as well as the restated program objectives contained in the same legislation ". . . to preserve and extend the service life of highways and enhance highway safety."

The background and origins of this controversy are discussed in greater detail in the remainder of this chapter. First, the evolution of federal involvement in funding highway construction is described, and the resulting federal-aid highway program is discussed. Discussed next is the division of responsibility between federal and state governments for setting highway design standards and the specific federal rulemaking efforts directed toward minimum geometric standards for RRR projects. Finally, a number of key issues related to the development and application of nationwide standards for RRR projects are summarized.

In Chapter 2 the following procedures are discussed: use of federal aid for RRR work by state and local highway agencies, project selection, type of projects undertaken, design standards and practices used, and the overall role of safety. Identified in Chapters 3 and 4 are the relationships between key geometric features and safety and those between key geometric features and cost. These relationships are identified as a preliminary step to the evaluation of the safety cost-effectiveness of geometric design standards. Wherever reliable quantitative relationships are identified, the safety and cost trade-offs of alternative standards are evaluated in Chapter 5, with an examination of project-level effects, as well as the implications for systemwide effects on safety, highway condition, and funding. The tort liability implications of geometric design standards and other RRR design practices are examined in Chapter 6. Finally, the study committee's key findings and recommendations for improved highway design practices are given in Chapter 7.

EVOLUTION OF FEDERAL HIGHWAY POLICY

Most highways in the United States are constructed, administered, and maintained by state and local governments. These tasks were accomplished with little federal assistance or involvement until the passage of the Federal-Aid Road Act of 1916. This act provided substantial financial assistance for highways, and in doing so, established the following key principles of federal highway policy:

- *State highway agencies:* specified that state highway departments would be the usual coordinator and contact point for all federal assistance;

- *Federal-state relationship:* affirmed the responsibility of state and local governments to construct, own, and maintain highways while committing the federal government to share in the financing of highway construction;
- *Federal-aid apportionment:* prescribed distribution of federal highway funds through a formula, which initially considered area, population, and rural postal route mileage;
- *Project cost sharing:* on federal-aid projects, required that federal funds be matched with state (or local) funds initially, with a maximum federal share of 50 percent;
- *Federal oversight:* specified that the federal government, originally through the Secretary of Agriculture, approve the plans, specifications, and estimates used for federal-aid highway projects *(1-3)*.

In succeeding legislation over the next 40 years, Congress continued to shape the federal-aid highway program, adding other major principles of federal policy.

- *Designated federal-aid highways:* In 1921 Congress directed that federal aid be limited to a designated system of interconnected highways. The system designated at the time was the forerunner of today's federal-aid primary system. Later, other legislation established additional designated federal-aid systems—the secondary system and urban primary extensions in 1934, the Interstate system in 1947, and the urban system in 1970.
- *User fee financing:* In 1932 the first federal excise tax on motor fuels was enacted. Although tax revenues were directed into the general fund, Congress in its deliberations began to link the tax rate to the level of federal highway expenditures.
- *Special categorical programs:* Through various highway and economic recovery legislation in the 1930s, Congress began to supplement federal aid for highway construction (regular federal aid) with categorical funds earmarked for specific purposes. The initial programs focused on eliminating the problems of railroad at-grade crossings, a federally funded activity that continues today.

By 1956 the basic principles of the federal-aid highway program were in place, but widespread concern existed over adequacy of U.S. highways, particularly in view of dramatic postwar increases in automobile and truck traffic. Through landmark legislation (the Federal-Aid and Highway Revenue Acts of 1956), the federal government responded to this concern, providing new funding to accelerate construction of the Interstate system and increased funding for other federal-aid systems. To finance this greatly expanded federal-aid program, the federal government increased existing highway-related excise taxes and established additional taxes whose revenues were totally or partly funneled into the Highway Trust Fund. (The trust fund was established as a holding mechanism for tax revenues earmarked for highway purposes.) The legislation required that the trust fund be used on a pay-as-you-go basis—effectively prohibiting federal deficit spending for highway construction.

From the outset financial assistance for the federal highway program (except for special categorical programs) was confined to relatively large, nonroutine projects—projects referred to in the Federal-Aid Road Act of 1916 as "substantial in character." In practice, this meant that federal aid was available only for constructing new roads or fully reconstructing existing roads to higher design standards. Lesser improvements to existing roads, including costly resurfacing or minor widening of roads constructed earlier with federal aid, were considered maintenance, and therefore were ineligible for federal aid until 1976.

Congress changed the distinction between construction and maintenance when it enacted the Federal-Aid Highway Act of 1976. In response to mounting concern over the condition of the nation's highway system and the need to shift emphasis from constructing new facilities to preserving existing ones, this act amended the U.S. Code to include resurfacing, restoration, and rehabilitation within the definition of "construction" as the term was used in the federal-aid highway program. The 1976 act authorized, for the first time, the use of federal funds for major repair work on the federal-aid highway system.

The act required that at least 20 percent of the regular federal aid for the primary and secondary systems be spent on RRR work. Later in the Surface Transportation Assistance Act of 1982, Congress modified this provision so that at least 40 percent of the primary, secondary, and urban system funds could be used on the combination of RRR work and reconstruction.

FEDERAL-AID HIGHWAY PROGRAM

The federal-aid highway program focuses on a designated system of highways and provides financial aid earmarked for particular components of this system or specific types of highway improvements. Congress authorized federal aid for RRR work within the context of the existing system of federal-aid highways and established funding categories.

Federal-Aid System

The designated federal-aid highway system accounts for 22 percent of the nation's total highway mileage but carries 81 percent of all traffic and accounts for 77 percent of all traffic fatalities (Table 1-1). It contains high proportions of the more heavily traveled expressways and arterial highways and few local roads and streets. For example, about 95 percent of all rural arterial and major collector highways are included in the federal-aid system.

TABLE 1-1 U.S. Highway Fatalities, Road Mileage, and Travel, 1985

	1985 Fatalities (thousands)		Mileage (thousands)		Vehicle-Miles Traveled (billions)	
	Number	Percent	Miles	Percent	Miles	Percent
Federal-aid systems						
Interstate	4.2	10	44	1	370	21
Primary	14.2	32	257	7	519	29
Secondary	6.5	15	398	10	156	9
Urban	8.9	20	144	4	383	22
Subtotal	33.8	77	843	22	1,428	81
Off-federal systems[a]	10.0	23	3,019	78	347	19
Total	43.8	100	3,862	100	1,775	100

SOURCE: FHWA, *Highway Statistics 1985 (4)* and *Highway Safety Performance, 1985 (5)*.
[a]Includes off-system local travel.

The federal-aid highway system comprises four components:

• *Interstate system:* Consists of about 44,000 mi of multilane expressways with fully controlled access traffic. The Interstate system is part of the federal-aid primary system but receives separate funding.
• *Primary system:* Excluding Interstates, consists of 257,000 mi of highways, of which 198,000 (77 percent) are two-lane rural roads.
• *Secondary system:* Initially designated in 1944, consists of 398,000 mi of highways classified as major collector roads in rural areas, nearly all of which (99 percent) are two-lane roads.
• *Urban system:* Includes 144,000 mi of urban arterial and collector roads, about 75 percent of which are two-lane roads.

This study is concerned with improvements to all systems except the Interstate, which in total account for about 95 percent of all federal-aid mileage, 60 percent of the nation's highway traffic, and two-thirds of its traffic fatalities. Two-lane rural highways alone account for about 75 percent of all nonfreeway federal-aid mileage and serve approximately one-fourth of vehicle miles traveled in the United States *(4)*.

The Federal Highway Administration maintains the Highway Performance Monitoring System (HPMS) that keeps track of the design characteristics and conditions of the U.S. highway system, including the federal-aid systems, based on a nationwide sample of highway segments. Using this system, FHWA (6) reported to Congress that approximately one-half of the federal-aid primary and secondary mileage has some type of significant geometric design deficiency related to either cross section (e.g., lane and shoulder widths) or alignment (e.g., curves, grades, and sight distance) (Table 1-2). About 16 percent of the mileage on these systems contains pavement deficiencies. In estimating these deficiencies, FHWA assumed "minimum tolerable conditions" that varied by functional classification, terrain, and traffic level.

TABLE 1-2 Estimated Deficiencies in Federal-Aid Systems (6)

	Federal-Aid System (Percentage of Paved Miles)			
	Interstate	Primary	Secondary	Urban
Pavement[a]	10.4	16.1	16.2	15.0
Horizontal and vertical alignment[b]	1.2	14.4	39.6	–
Cross section[c]	4.6	45.9	38.5	34.2
Operational[d]	11.0	19.2	3.7	14.3
Access control[e]	0.1	0.1	–	0
Miles with single deficiency	17.8	41.5	40.1	40.7
Miles with more than one deficiency	4.5	23.9	26.5	11.2
Total miles with deficiency	22.3	65.5	66.6	51.9

NOTE: Estimates based on HPMS data for 1981.
[a]Pavement condition is based on the Present Serviceability Rating (PSR). A pavement with a PSR of 2.0 or lower (2.5 for Interstates) is considered deficient.
[b]Highway segments with curves that require reduced speed, or any grades with insufficient sight distance that require trucks to slow on arterials and major collectors, or more than an occasional such curve or grade on minor collectors.
[c]Lane widths less than 11 ft on principal arterials, 10 ft on minor arterials and major collectors, or 8 to 9 ft on minor collectors; or shoulder widths on arterials and major collectors of less than 4 to 8 ft, depending on terrain and traffic.
[d]Operational deficiencies in rural areas occur when operating speed drops below a threshold that is a function of classification, average daily traffic (ADT), and terrain. In urban areas the definition is based on the peak period volume-to-capacity ratio.
[e]A deficiency results when a segment of highway that should be access controlled is not; this applies to all Interstates and certain primary and urban routes.

Comparatively, the Interstate system contains fewer deficiencies owing to its more recent design and construction, as well as the higher maintenance priority it generally receives. The FHWA estimated that less than 10 percent of Interstate mileage contains geometric deficiencies and approximately 10 percent of its pavement is deficient.

The federal-aid systems include approximately 267,000 bridges, which account for about one-half of all U.S. bridges. The FHWA (7) estimates that

23 percent of these (more than 60,000 bridges) are eligible for special federal bridge replacement and rehabilitation funds because of existing deficiencies in design and condition. About 80 percent of these bridges are on either the primary or secondary federal-aid system.

Funding Programs

The overall federal-aid highway program[1] consists of more than 40 separate funding assistance programs (8). However, the construction programs for the designated federal-aid systems (regular federal aid plus minimum state allocations) account for about 80 percent of all federal assistance (Table 1-3). In general, these funds are apportioned to the states by formulas that vary depending on the system.

Although most highway programs have some direct or indirect effect on safety, over the years Congress has established three programs that fund safety

TABLE 1-3 Expenditure of Federal-Aid Highway Funds Administered by the Federal Highway Administration During 1985

	Expenditures Paid from the Highway Trust Fund	
	($ millions)	(percent)
Primary	2,239	17.4
Secondary	590	4.6
Urban	833	6.5
Interstate	3,923	30.4
Interstate resurfacing	2,322	18.0
85 percent minimum allocation funds	316	2.5
Planning and research	202	1.6
Highway safety	423	3.3
Bridge replacement	1,511	11.7
Other	525	4.1
Total	12,884	100.0

SOURCE: *Highway Statistics 1985 (4)*, Table FA-3.

---

[1] This section contains a description of federal funding programs as authorized through fiscal year 1986. Shortly before this writing (in April 1987), the Surface Transportation and Uniform Relocation Assistance Act of 1987 became law. This act authorizes expenditures on federal-aid highway projects and safety programs for fiscal years 1987 through 1991. Although the act provides some changes in funding priorities, it does not substantially change those elements of the federal-aid highway program that are of interest in this study.

improvements at specific locations, often involving construction or rehabilitation:

• *Bridge replacement and rehabilitation:* Since 1979 this program has funded bridge replacement and rehabilitation projects on and off the federal-aid system. Before the 1978 Highway Act, bridges on the federal-aid system could be partially replaced (e.g., new deck) under the special bridge replacement program.

• *Rail-highway grade crossings:* Established by the Highway Safety Act of 1973, this program provides funds for safety improvements at on- and off-system rail crossings, including grade separation, relocations, automatic gates, and warning devices.

• *Hazard elimination:* Since 1974 this program and its predecessors have funded on- and off-system spot safety improvements, frequently at locations with histories of high accident rates. Through fiscal year 1982, program funds were directed to improvements such as traffic signals and intersection channelization (34 percent), guardrail installation (15 percent), signs and breakaway supports (2 percent), and pavement skid treatments (4 percent) *(9)*.

Just as it has done with regular federal-aid funds, Congress has authorized some transferability between the bridge replacement and rehabilitation, rail-highway grade crossings, and hazard elimination categorical programs.

In recent years federal aid from all programs has accounted for approximately 25 to 30 percent of total highway disbursements ($54 billion in 1985) at all levels of government. During the years when major Interstate construction was underway, capital outlays, which include most federal aid, represented more than one-half of total disbursements; but in recent years, capital outlays have shrunk to less than one-half of the total while maintenance and operational disbursements have increased as a percentage. Moreover, the buying power of capital funds decreased dramatically during periods of rapid inflation in the 1970s so that, despite recent improvements in construction buying power, the FHWA *(5)* reported that 1982 capital outlays were expected to be less in constant dollars than any year since the early 1950s. This trend was reversed, at least temporarily, by the Surface Transportation Assistance Act of 1982, which increased 1983 federal-aid highway system authorization levels about 40 percent over the 1982 level. Further, it authorized annual increases in the Interstate resurfacing, restoration, rehabilitation, and reconstruction (4R) and primary system categories each year between 1983 and 1986. Fiscal year 1986 Interstate 4R authorizations were nearly four times the 1982 level, and primary authorizations were nearly twice the 1982 amount.

Resurfacing, Restoration, and Rehabilitation

Under the classification of resurfacing, restoration, and rehabilitation, the federal-aid highway program funds the following types of improvements to existing federal-aid highways: resurfacing, pavement structural and joint repair, minor lane and shoulder widening, minor alterations to vertical grades and horizontal curves, bridge repair, and removal or protection of roadside obstacles. In making RRR work eligible for regular federal aid, Congress made federal funds available for the heavier, more costly types of maintenance and at the same time provided highway agencies the opportunity to use federal funds for incremental geometric and safety improvements short of full reconstruction. RRR is not a separate federal-aid program but rather a collection of improvement types that became eligible for regular federal aid in 1976. Currently, in distinguishing between RRR work and ineligible routine maintenance, FHWA classifies as maintenance any overlays less than ¾ in., pavement repairs on short segments, and patching and repair of minor pavement failures (10).

Between 1977 and 1982, when federal law required at least 20 percent of regular primary and secondary funds be used for RRR, FHWA also had to distinguish between RRR work and reconstruction so that compliance could be verified. In general, FHWA views the complete removal and replacement of pavement structure or the addition of new continuous traffic lanes as reconstruction, rather than RRR (11). After 1983, when the law changed to require at least 40 percent of non-Interstate regular federal aid be used for the combination of RRR and reconstruction, the distinction between the two became less important because the treatment was identical in the revised program.

Not all federal-aid RRR work is funded with regular federal-aid construction funds. Some RRR-type projects may be funded with either regular federal aid or separate categorical aid. For example, state or local highway agencies might seek federal funds for a bridge rehabilitation project through either regular federal aid (as an eligible RRR project) or through the bridge replacement and rehabilitation program. Similarly, roadside hazard removal and guardrail installation might, in some circumstances, be funded with either regular federal aid or hazard elimination program funds.

State highway agencies spent $2.9 billion on RRR projects in 1985, including $1.7 billion federal-aid and state matching funds and $1.2 billion state funds (4). Counties and municipalities spent approximately $2.1 billion on

RRR projects.[2] Total spending on RRR projects in 1985 was $5.0 billion, about 10 percent of expenditures for highways by all levels of government combined.

Many state highway agencies use RRR federal aid as a means of addressing critical pavement preservation and repair needs while making selective improvements to road geometry and roadside features. For example, a state might use federal aid for a resurfacing project that would also widen the shoulder and replace obsolete guardrail. The extent of such geometric and roadside improvements varies among the states and is at the center of the controversy that has existed regarding geometric design standards for federal-aid RRR projects.

GEOMETRIC DESIGN STANDARDS AND FEDERAL RULEMAKING

Responsibility for Standards—Federal and State Roles

Historically, the primary responsibility for developing and adopting design standards has rested with state legislatures, state highway agencies, and local governments. These standards generally set minimum values for key geometric features and sometimes call for specific design procedures or practices as well. Over the years highway design standards have been developed principally for the construction of new highways or the complete reconstruction of existing highways.

With the passage of the Federal-Aid Road Act of 1916, the federal government became involved with highway design standards for the first time. Initially, it exercised oversight by approving designs for federal-aid projects on a project-by-project basis. It did not develop or seek to apply nationwide standards but did begin sponsoring early research on geometric design, the results of which were later incorporated in state standards and American Association of State Highway and Transportation Officials (AASHTO) design policies. For example, in 1925 the Bureau of Public Roads, predecessor of FHWA, reported that 18 ft was the minimum pavement width for trucks and automobiles to pass safely; later in 1944 a federal study recommended a lane width of 12 ft *(1)*.

---

[2]RRR accounted for 47 and 26 percent, respectively, of capital outlay for state-funded projects on rural and urban collectors. These percentages were applied to total capital outlays by counties and municipalities ($2.4 and $3.6 billion, respectively) to estimate their RRR expenditures, yielding $1.1 billion for counties and $1.0 billion for municipalities.

Until the late 1920s, state agencies generally adopted standards independently of one another, leading to design inconsistencies between adjacent states as well as duplications of effort. To address these problems, the American Association of State Highway Officials (AASHO) (now the American Association of State Highway and Transportation Officials), which initially confined itself to disseminating information, began to adopt design policies in 1928 *(12)*. These policies, though not obligatory, were intended by AASHO to guide its members on technical matters in which state-to-state uniformity was needed. By 1944 it had adopted seven design policies that were incorporated into the design standards of many states *(13)*.

Over the years, AASHTO has revised these design policies a number of times and issued many additional policies *(14)* that not only recommend minimum design values but also procedures to be used in the planning and design of highways. For example, AASHTO has recommended procedures for citizen participation, environmental studies, and project evaluation *(15)* and has issued guidance materials for designing pavements *(16)*, traffic barriers *(17)*, and lighting *(18)*.

The federal government adopted many of AASHTO's design policies and guides as standards for federal-aid projects; approximately 20 are incorporated into federal regulations of highway design *(19)*. Although FHWA has supplemented AASHTO design policies in some areas, it has generally preferred to adopt them rather than develop design standards independently and has participated in AASHTO committees responsible for developing design policies. In practice, each state highway agency continues to incorporate AASHTO policies in its design standards, often with some alterations or extensions. FHWA reviews and must approve these standards for use on federal-aid projects.

For road geometry, AASHTO policies recommend minimum (or maximum) design values for features such as

- Lane widths,
- Shoulder widths,
- Horizontal and vertical curves,
- Superelevation at curves,
- Sight distance,
- Bridge widths,
- Sideslopes and ditch drainage, and
- Pavement cross slopes.

To increase flexibility and adaptability to a variety of nationwide conditions, AASHTO recommends different design values for variations in terrain, setting (urban versus rural), traffic volume, traffic characteristics (e.g., percentage of heavy trucks), and function (local, collector, arterial, etc.). The recommended values also vary with speed. For roads intended for high-speed driving, the values specify wider lanes and shoulders, longer sight distances, and more gentle curves. AASHTO policies recommend design speeds based on function, setting, terrain, and traffic characteristics.

Not all highway features are treated with firm, numerical design standards. Roadside features such as obstacle-free clear zones and protective barriers generally have been covered by guidelines rather than recommended minimum design values. As a result, new construction design is more uniform on the roadway than on the adjacent roadside.

Geometric Standards for RRR Projects

Geometric standards for RRR projects specify whether particular geometric features must be upgraded as part of the project. Features that do not meet minimum standards must be upgraded unless a design exception is sought and approved.

When Congress authorized federal aid for RRR projects in 1976, AASHTO had not developed design policies specifically for RRR work, nor had FHWA adopted minimum geometric standards for RRR projects. Existing standards and policies were geared to new construction or reconstruction. As an interim measure, the FHWA applied its new construction standards to RRR projects while it considered separate RRR standards for nonfreeway highways. For freeways, mostly Interstate highways constructed in the past 30 years, FHWA concluded that no special RRR standards were warranted because these highways generally met the most stringent new construction standards.

Design exceptions, permitted on a case-by-case basis for any federal-aid project quickly became commonplace for nonfreeway, federal-aid RRR projects because upgrading to the geometric standards for new highways is often extraordinarily expensive. In the northeastern states, where highway systems are relatively old and the topography is severe, the FHWA reported that 75 to 90 percent of RRR projects were granted design exceptions. In those middle and far west states where highways are newer and the topography is relatively flat, the percentage of RRR projects granted design exceptions was far lower (0 to 30 percent) *(20)*. Differences in highway widths contribute to the higher percentage of projects in the northeast granted design exceptions. For example, in all but one state west of the Mississippi more than 60 percent of the primary highways now have lanes 12 ft wide, whereas few primary highways in northeastern states have 60 percent of primary highways with lanes of this width *(4)*.

FHWA noted that not all of the differences among states in the percentages of RRR projects granted design exceptions can be explained by differences in topography and the age of highway systems *(20)*. As discussed in more detail in Chapter 2, other factors that affect these percentages include the amount of state funds available for RRR, the design standards in effect, and the procedures followed by FHWA division offices in approving design exceptions.

By November 1976, AASHTO had developed and adopted a policy on geometric design for RRR work, which was published in the RRR geometric design guide the following year *(21)*. Commonly referred to as the "purple book," this guide contains recommended minimum design values for lane and shoulder widths, cross slopes, superelevation, and bridge widths, as well as advisory information on grades, curvature, sight distance, and clear zones. Overall, it is considerably less stringent than AASHTO policies for new construction (Appendix A). The purple book was opposed by safety organizations, and within FHWA it was opposed by the Office of Highway Safety.

The FHWA's interim measure of requiring new construction standards for RRR projects, with lenient exceptions, lasted far longer than expected because selecting separate RRR geometric design standards for nonfreeway highways proved to be complex and controversial. Safety and design issues were raised that involved a large number of geometric features affecting different types of highways in rural and urban settings. Underlying the question of standards was the need to use federal RRR funds in the most cost-effective manner to enhance safety while preserving and restoring federal-aid highways for the nation as a whole. Controversy was introduced by the different perspectives of safety organizations and state highway agencies.

It was not until June 1982 that the FHWA issued new regulations addressing RRR standards. During the intervening 6 years, the FHWA considered a number of alternative policies and reversed itself with respect to preferred action. The following alternatives were considered:

1. Continue to use new construction geometric design standards with exceptions permitted on a case-by-case basis.
2. Adopt guidelines contained in the AASHTO RRR design guide.
3. Adopt RRR standards developed by the FHWA. In August 1978, after opposition to the AASHTO RRR guidelines arose, the FHWA proposed RRR standards developed internally *(22)*. In general, the FHWA standards are somewhat more stringent than the standards in the AASHTO RRR guidelines, but are similar in terms of scope and format (Appendix A, Table A-1).
4. Adopt a flexible approach under which states could develop and use their own RRR standards subject to FHWA approval.

This protracted rulemaking process attracted comments from a variety of institutions and individuals, including safety-oriented organizations such as the Center for Auto Safety, the Insurance Institute for Highway Safety, and the

National Transportation Safety Board. Safety organizations generally opposed any regulation that might lead to special standards for RRR projects and favored the first alternative as least objectionable.

Although the FHWA had granted a large number of design exceptions under the first alternative, safety organizations believed the process of explicitly considering design exceptions on a project-by-project basis will occasionally result in substantial geometric improvements. Safety organizations acknowledged the need for design exceptions, but they viewed the AASHTO RRR guidelines and FHWA proposed standards as too lax, permitting the RRR program to focus almost exclusively on road surface improvements and discouraging a safety-conscious design process. In addition, it was argued that reductions in standards for federally assisted RRR projects would violate legislative mandates concerning safety. The fourth alternative, permitting states to develop their own standards, also was unpopular with safety organizations, which feared that states would choose, and the FHWA would approve, standards similar to the AASHTO RRR guidelines *(23)*.

State highway agencies initially supported the AASHTO RRR guidelines, but later indicated a general willingness to accept the more stringent RRR standards proposed by the FHWA. New construction standards were viewed by many states as inappropriate for RRR projects. State officials generally believed that the new standards, if followed rigorously, would greatly increase project costs, thereby concentrating available funds on a small number of improvement projects. Such a policy, it was argued, would leave unattended many miles of federal-aid highways in need of pavement repair and would meet neither safety nor repair objectives. It was also contended that if widespread exceptions were permitted, needless administrative costs and delays would be incurred *(23)*.

In June 1982 the FHWA selected the fourth approach, permitting states to develop their own RRR standards subject to FHWA approval. By this time, some states had grown accustomed to using new construction standards, with case-by-case exceptions, and under the June 1982 rule, states were permitted to continue this practice *(9)*. However, the issue remained unsettled because of congressional reservations concerning this approach. These reservations initially surfaced during the fall of 1981 when FHWA officials, representatives of safety organizations, state highway officials, and others debated RRR issues extensively in hearings held by the House Subcommittee on Investigations and Oversight of the Committee on Public Works and Transportation *(24)*.

Safety Considerations

In the congressional hearings on RRR standards, debate focused on *(a)* the practical impact of RRR standards on safety and *(b)* the role of safety in the

RRR program. A number of questions were raised with respect to the impact of the standards on safety. Will accident rates increase if highways with existing geometric deficiencies are resurfaced and no other improvements are made? What changes in accident rates can be expected if different types of geometric improvements are made? What would be the nationwide consequences on safety and highway condition of alternative RRR standards when budgetary resources are fixed? What would be the biggest safety payoff? Although such questions could not be answered fully, the FHWA prepared the RRR Technical Analysis report to address them. This analysis, which concluded that standards less stringent than those for new construction would be appropriate for RRR projects *(1)*, was criticized by the National Transportation Safety Board for methodological shortcomings *(25)*. Overall, many issues concerning the desired level of safety to be included in the program, and the type of standards and policies needed to balance safety and pavement preservation, were left unresolved during the RRR rulemaking process.

With respect to the second issue, the role of safety in the RRR program, the FHWA adopted the position that safety was an essential consideration of the RRR program, but secondary to preserving and extending the service life of highways. On the other hand, safety organizations expressed the concern that efforts to upgrade the safety of federal-aid roads were being relaxed in the RRR program. Debate over the relative priority of safety and repair arose repeatedly in the testimony during the congressional RRR hearings *(23)*.

Ultimately, this debate led to a provision in the Surface Transportation Assistance Act of 1982, which stated that the objective of the RRR program is ". . . to preserve and extend the service life of highways and enhance highway safety." Congressional deliberations were unclear about how much of a change, if any, was required by this provision *(26)*. Subsequently, the FHWA modified its June 1982 rule on RRR standards to echo this restatement of program objectives. Reflecting the legislative ambiguity, the FHWA changed the policy statement in the preamble to the rule, but made no changes to the procedures *(27)*.

These changes to statutory language and rules have not resolved the problem of the cost-safety trade-off within the RRR program. To address "safety cost-effectiveness," an additional provision was included in the Surface Transportation Assistance Act of 1982 that called for the National Academy of Sciences to study the safety cost-effectiveness of highway geometric design standards and recommend minimum standards for resurfacing, restoration, and rehabilitation projects on existing federal-aid highways, except freeways.

KEY ISSUES

Many questions and issues that bear on minimum RRR standards were left unresolved during the RRR rulemaking process and related debate. In organizing a study that would respond to the congressional request, the study identified six key areas of inquiry that address these issues and, taken together, provide the technical foundation necessary for specific recommendations.

State and Local Procedures for Selection, Design, and Construction of Highway Improvement Projects (Chapter 2)

To analyze alternative design standards, it is necessary to understand the relationship between standards and other parts of the RRR process: What types of RRR projects are funded with federal aid? How are these projects selected? What design standards are currently used? Are stringent design standards frequently circumvented? How are safety needs taken into account? All of these questions are directed to state and local highway agencies because they have the primary responsibility for selecting and performing RRR work.

During the congressional hearings on RRR standards (24), witnesses relied heavily on either nationwide statistical data or personal observations about specific state and local practices. Although useful, nationwide statistical data or studies may mask significant variations among the states with respect to RRR needs and the ways by which they are met. Personal observations may be accurate but are difficult to compare systematically. Further, they may portray an atypical situation reflecting the practices of a single highway agency.

To obtain a more balanced picture, reviews of prior studies and analyses of nationwide data bases were supplemented with in-depth case studies of highway practices in 15 states and interviews with local public works officials throughout the country.

Relationship Between Safety and Geometric Design (Chapter 3)

What are the safety payoffs (i.e., reductions in the number and severity of accidents) from geometric changes such as increasing lane and shoulder widths, straightening sharp curves, or removing roadside obstacles on existing highways? The trade-off between the cost of such improvements and their safety payoff is fundamental to the issue of minimum RRR design standards. To make this trade-off requires quantitative knowledge of the relationships between safety and different roadway features. Despite numerous statistical studies of accident data, these relationships are not well known, and divergent relationships are suggested by different analyses *(28)*. Isolating the effects of a specific geometric feature from other conditions of the roadway environment, vehicle characteristics, and driver characteristics has proved to be a formidable research task. This task is often complicated by the lack of comprehensive and consistent accident and exposure (usage) data.

Highway researchers will probably never develop definitive safety relationships that cover the full range of highway design features. The complexity of the causes of accidents, the infrequency of accidents, and the continuing evolution of highway vehicles, traffic regulations, and enforcement policies work against this. Nevertheless, wherever possible, the study committee made judgments about the most probable relationships between safety and key highway features using what was considered to be the most credible data available. To help provide a basis for these judgments, the committee sponsored critical reviews of prior research and two special studies that addressed major gaps in existing knowledge.

Relationship Between Cost and Geometric Design (Chapter 4)

Like the relationships between safety and geometric design, the relationships between cost and geometric design are critical in determining how safety and road repair needs can most effectively be balanced. How much does it cost, for example, to widen lanes and shoulders in addition to resurfacing as opposed to simply resurfacing? Nationwide statistics compiled by FHWA for federal-aid projects indicate that typical resurfacing projects on rural arterials cost approximately $150,000 to $200,000/mi. When minor widening is included, the cost more than doubles. Full reconstruction, with wide lanes, costs approximately $1.25 million/mi, more than six times the unit cost of simple resurfacing *(29)*.

Although these rough estimates provide a sense of the magnitude of costs involved, they mask the large variability that exists from region to region, state to state, and from project to project.

To develop cost relationships for use in the study, the committee examined published cost data, cost records, and estimating procedures for a sample of state highway agencies throughout the country. This work supplemented reviews of existing nationwide cost records and data sets.

Safety Cost-Effectiveness of Geometric Design Standards (Chapter 5)

The principal questions that underlie earlier debate over RRR standards are

- What are the cost and safety trade-offs of making incremental geometric improvements to existing highways?
- How do minimum RRR standards affect the balance between preserving highways and improving safety on a systemwide basis, when available funds are limited?

These questions were addressed from project-level and system-level perspectives using the safety and cost relationships identified. The added cost per accident eliminated that can typically be expected for improvements to existing highway geometry was estimated at the project level. How much does it cost, for example, to eliminate an accident by lane widening, and how does this compare with shoulder widening or straightening a sharp horizontal curve? Where system data were available for existing highway conditions, on a nationwide basis and for selected states, the study estimated the effect of alternative RRR minimum standards on systemwide safety and the total expenditure needed to meet the standards. Also considered was the likely impact that such standards would have on the frequency of major pavement repairs and operational benefits in the form of reduced user costs that may result from geometric improvements along with improved safety.

Tort Liability and Geometric Design (Chapter 6)

Highway agencies have become increasingly concerned about the number of tort claims filed against them and the resulting costs of settlements, awards, and legal defense. These claims allege negligence in the design or operation of public highways.

Some highway agencies have feared that special geometric design standards for RRR projects, less stringent than new construction standards, might make them more susceptible to tort claims; others have concluded just the opposite. The limited data available on tort claims against highway agencies were analyzed to determine how frequently geometric design is at issue as opposed to maintenance practices, signing, or other aspects of highway management. In addition to standards, other ways were considered in which the design and construction of RRR projects may reduce a highway agency's exposure to tort claims.

REFERENCES

1. *America's Highway's 1776-1976: A History of the Federal-Aid Program.* FHWA, U.S. Department of Transportation, 1976.
2. P. K. Wheeler. *Highway Assistance Programs: A Historical Perspective.* Congressional Budget Office, Washington, D.C., Jan. 1978.

BACKGROUND AND ISSUES   33

3. Federal-Aid Road Act of July 11, 1916, Ch.241, 30 Stat. 355.
4. *Highway Statistics 1985*. FHWA, U.S. Department of Transportation, 1986.
5. *Highway Safety Performance—1985, Fatal and Injury Accident Rates on Public Roads in the United States*. FHWA, U.S. Department of Transportation, (forthcoming).
6. *Status of the Nation's Highways: Conditions and Performance*. FHWA, U.S. Department of Transportation, June 1983.
7. *Fifth Annual Report to Congress: Highway Bridge Replacement and Rehabilitation Program*. FHWA, U.S. Department of Transportation, May 1983.
8. *Financing Federal-Aid Highways*. FHWA, U.S. Department of Transportation, Sept. 1983.
9. *The 1986 Annual Report on Highway Safety Improvement Programs*. FHWA. U.S. Department of Transportation, April 1986.
10. *Resurfacing, Restoration, and Rehabilitation (R-R-R) Work*. FHWA Notice N 5040.19. U.S. Department of Transportation, June 28, 1976.
11. "Design Standards for Highways," final rule. *Federal Register*, Vol. 47, No. 112, June 10, 1982, pp. 25268-25275 (FHWA project coding definition).
12. F. W. Cron. "Highway Design for Motor Vehicles—A Historical Review." Part 8: "The Evolution of Highway Standards." *Public Roads*, Vol. 40, No. 3, Dec. 1976, pp. 93–100.
13. *A Policy on Geometric Design of Rural Highways: 1965*. American Association of State Highway Officials, Washington, D.C., 1966.
14. *A Policy on Geometric Design of Highways and Streets*, 1984. American Association of State Highway and Transportation Officials, Washington, D.C.
15. *A Policy on Design of Urban Highways and Arterial Streets*, 1973. American Association of State Highway and Transportation Officials, Washington, D.C.
16. *Interim Guide for Design of Pavement Structures*. American Association of State Highway and Transportation Officials, Washington, D.C., 1972.
17. *Guide for Selecting, Locating, and Designing Traffic Barriers*. American Association of State Highway and Transportation Officials, Washington, D.C., 1977.
18. *An Information Guide for Roadway Lighting*. American Association of State Highway and Transportation Officials, Washington, D.C., 1976.
19. Design Standards for Highways, Code of Federal Regulations, Title 23, Part 625.
20. U.S. Congress. House of Representatives. Testimony of L. Lamm. Subcommittee on Investigation and Oversight, Committee on Public Works and Transportation, *Resurfacing, Restoration, and Rehabilitation of Roads Other Than Freeways*. Hearings . . . . , 97th Congress, Sept. 17, 1981 (Serial No. 97–75, p. 45).
21. *Geometric Design Guide for Resurfacing, Restoration, and Rehabilitation (R-R-R) of Highways and Streets*. American Association of State Highway and Transportation Officials, Washington, D.C., 1977.
22. "Design Standards for Highways." Notice of proposed rulemaking. *Federal Register*, Vol. 43, No. 164, Aug. 23, 1978, pp. 37556–37568.
23. Unpublished summaries of comments to rulemaking Dockets No. 77–4 and No. 77–10, FHWA, U.S. Department of Transportation (undated).

24. U.S. Congress. House of Representatives. Subcommittee on Investigations and Oversight, Committee on Public Works. *Resurfacing, Restoration, and Rehabilitation of Roads Other Than Highways,* Hearings . . . 97th Congress, Sept. 17, Oct. 27, 28; Dec.15, 1981.
25. *Federal Highway Administration Non-Interstate Resurfacing, Restoration, and Rehabilitation Program: Safety Effectiveness Evaluation.* National Transportation Safety Board, Washington, D.C. Sept. 22, 1981.
26. House (p. H10717, December 21, 1982) and Senate (p. S16067, December 23, 1982) colloquies to the effect that the "enhance safety" provision would not require the application of full new design standards.
27. "Design Standards for Highways—Resurfacing, Restoration, and Rehabilitation of Streets and Highways other than Freeways." Final rule. *Federal Register*, Vol. 48, No. 63, March 31, 1983, pp. 13410–13412.
28. *Synthesis of Safety Research Related to Traffic Control and Roadway Elements.* FHWA, U.S. Department of Transportation, Vol. 1, Dec. 1982 (see Chapter 4, "Roadway Cross Section and Alignment").
29. *State of the Nation's Highways: Conditions and Performance.* (Table A–4, p. A–15). FHWA, U.S. Department of Transportation, June 1983.

2
State and Local Procedures for Selection, Design, and Construction of Highway Improvement Projects

State and local (county and municipal) highway agencies are responsible for constructing and maintaining the nation's federal-aid highways. These agencies select and design resurfacing, restoration, and rehabilitation (RRR) projects as part of this responsibility. The federal role is limited to providing funding assistance and exercising oversight authority, largely by setting design standards and guidelines, determining project funding eligibility, and approving individual project designs. Because of this division of responsibilities, the effectiveness of any federally imposed requirement, such as RRR design standards, cannot be easily judged or evaluated without also examining parts of the overall process controlled by state and local highway agencies. Nor can changes in federal RRR requirements be reasonably assessed without first understanding the context in which they operate and how they interact with other steps in the process to promote preservation and enhance highway safety—the statutory objectives of RRR conducted with federal aid.

This chapter contains the results of a review by the study committee of resurfacing, restoration, and rehabilitation activities in state highway agencies and local governments. The review was conducted to

- Determine the nature of state and local RRR projects—their characteristics and objectives in the context of state and local agencies' overall construction and maintenance programs;

- Develop an understanding of the process in state and local agencies that determines the characteristics of RRR projects. This process includes not only project design (application of RRR standards, but also programming (selecting and scheduling RRR projects) and finance (determining the overall level of resources to be devoted to RRR and the sources of funding); and
- Focus the study committee's evaluation of RRR design standards and practices on the critical areas of uncertainty or disagreement, and identify possible alternatives to current practices that may be valuable in meeting federal-aid RRR program objectives.

The characteristics of RRR projects were considered relevant to the committee's review because *(a)* RRR standards and guidelines must be appropriate for the projects to which they are applied, *(b)* project characteristics determine which standards are likely to be binding, and *(c)* the nature of the projects reflects the objectives of state and local highway agencies in the use of federal aid for RRR. The following specific questions about RRR project characteristics are addressed in this chapter: What are the immediate purposes of state and local federal-aid RRR projects—are they mainly resurfacing and other pavement repairs, safety enhancement, a general upgrading of the level of service, or a combination of these? What type of geometric or other safety-motivated improvements are commonly made as part of resurfacing projects? How do federal-aid projects differ from non-federal-aid projects?

Project selection, programming, and financing are important because decisions in these areas critically affect the success of the federal-aid RRR program in meeting its objectives. In the discussion of RRR programming and finance the following questions are addressed: How do state and local agencies select RRR projects? What factors (pavement condition, geometric features, accident history, traffic, or others) do they consider? Do federal RRR standards influence project selections? How do financial constraints such as a shortage of state capital improvement funds or funding allocation procedures imposed by a state legislature affect these choices? How does the state choose between federal and state funding for RRR?

Design standards and practices determine the nature and quality of resulting projects. The following questions relating to project design are addressed in this chapter: What design standards are used? How are safety needs and objectives considered? What is the Federal Highway Administration (FHWA) role in influencing the states' design practices?

In addition to RRR practices, the committee also examined the relationship between RRR activities and other highway programs with similar objectives, including the special federal-aid programs supporting bridge replacement and rehabilitation, hazard elimination, and state-funded pavement maintenance activities.

REVIEW OF RRR PRACTICES: INFORMATION SOURCES

To answer the preceding questions, the committee first reviewed existing studies and data that address procedures for developing federal-aid RRR projects or describe resulting projects. Relevant previous studies are principally an FHWA review of RRR projects in 19 states *(1)* and two General Accounting Office (GAO) projects—one directed at the nature of RRR design exceptions granted by FHWA *(2)*, and another aimed at determining the type of federal-aid RRR and reconstruction work performed *(3)*. The committee also examined nationwide data maintained by FHWA to record the status, work classification, and basic financial information on all federal-aid projects.

Information available from these sources was expanded and updated through case studies of RRR programs in 15 state highway agencies and interviews with more than 40 local highway officials familiar with RRR activities. The case study states were selected for the diversity of factors that may influence RRR needs and programs: highway age and condition, terrain, mix of urban and rural conditions in the state, agency organization, size of federal-aid apportionments, availability of state and local funds to supplement federal aid, and RRR design standards used. (The case study states, along with local highway agencies and their RRR activities, are described in Appendix B, Tables B-1–B-15.) The 15 states selected account for nearly 50 percent of federal-aid apportionments and include most of the states with the largest highway programs.

Committee staff visited the highway agency and the FHWA office in each of the 15 states and reported to the committee RRR activities and the way these activities fit into the overall state highway program.

Of the 15 state highway agencies visited, 9 were using special RRR design standards approved by FHWA and 6 had special agreements with FHWA exempting them from many routine FHWA reviews of project designs (an arrangement known as certification acceptance) (Appendix B, Table B-1).

Telephone interviews were conducted with local highway officials representing 16 counties, 20 cities, and 3 metropolitan planning organizations to discuss RRR practices. Most of the officials were selected from jurisdictions in the case study states through state highway agency referrals or independent contacts with state or national associations of local officials.

Interviews with state and local highway agency officials were conducted during 1984. Follow-up interviews conducted in the fall of 1986 indicated that few important changes occurred in the states between 1984 and 1986. The conditions that existed in 1984 are reported in this chapter and the tables in Appendix B; the summaries of RRR standards in Tables B-9 and B-10 have been updated to reflect conditions in 1986. Other changes since 1984 are summarized in the introduction to Appendix B.

STATE RRR PROGRAMS

This section contains background information on the organization of state highway agencies and an examination of typical state RRR project characteristics, programming procedures, financing, and design standards and practices.

State Highway Agency Organization

Functions

The principal functions of state highway agencies are

- Construction;
- Project development and design;
- Maintenance;
- Traffic operations;
- Programming and budgeting;
- Local government financial assistance; and
- Support, including general administration, research and testing, and long-range planning and data collection.

State highway agencies are responsible for the major intercity routes in the state, including federal-aid Interstates and most federal-aid primaries, but share responsibility for administering federal-aid secondary and urban roads with local governments. The extent of this shared responsibility varies widely, particulary for federal-aid secondary roads (Appendix B, Table B-2). In Missouri the state administers all federal-aid secondary roads; however, in New Jersey the state administers only 5 percent of such roads. Among the case study states, the share of total federal-aid mileage administered by the state ranges from 94 percent in Missouri to 26 percent in New Jersey. In most of the case study states, cities and counties maintain nearly all local rural roads and city streets, which are ineligible for most federal-aid programs, but a few state highway agencies (Virginia, New Hampshire, Texas, and Missouri) administer substantial non-federal-aid rural mileage.

Organizational Structure

In most state highway agencies, responsibilities are divided between a central office and several district offices. The central office is organized into functionally defined divisions (e.g., design, construction, maintenance) whereas district offices have geographically defined responsibilities for maintenance and other functions as well. Larger state highway agencies tend to be more decentralized (e.g., New York, California, Texas) with significant responsibility for project programming and design handled by the districts and policy and review functions performed by the central offices. Smaller states are more likely to have highly centralized organizations (e.g., South Dakota, New Hampshire) with the central office responsible for all functions except maintenance and supervision of construction.

Highway agency organizational charts differ from state to state, but there are usually parallel divisions for the major functional areas of project development and design, construction, maintenance and traffic operations, planning, and administrative services (Figure 2-1).

Assignment of project programming responsibilities varies most in the organizational structure of highway agencies; the programming office may be under planning, project development, or administration, or it may be a separate division. The programming office is responsible for assembling a multiyear construction program (listing projects, costs, and funding sources) based on program proposals from the functional divisions and the district offices, matching projects to funding sources, ensuring that the program meets all departmental and legislative requirements regarding funding allocations to program categories and geographic areas, and ensuring that all federal-aid funds for which the state is eligible are used. The programming office may also have substantial authority for final project selection and scheduling decisions, especially for smaller projects. Top management also is active in the selection of all major projects and in final approval of the program.

In decentralized highway departments, the major activity of a central office functional division is to oversee district activities in its area of responsibility. For example, the central office design division will develop standards that district designers must follow, approve project designs for the districts, and periodically review district performance.

RRR Responsibilities

In the past, most state highway agencies performed resurfacing as a part of their maintenance programs. Historically, maintenance activities were funded almost entirely by the state (without federal assistance), but this situation

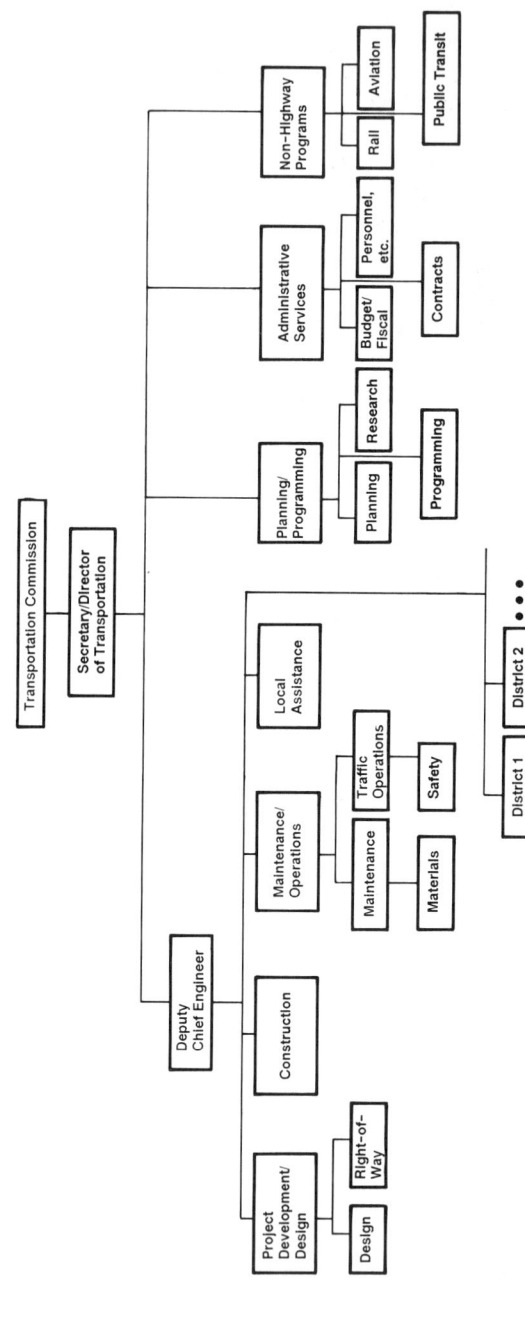

FIGURE 2-1 Typical state highway agency organization.

changed in 1976 when Congress made RRR eligible for federal aid. Some states responded by using federal aid for projects conducted under their maintenance programs. However, this practice created problems—maintenance staffs were not familiar with the administrative aspects of federal aid, and, moreover, were oriented toward pavement repair projects that did not require extensive design plans or involve the concurrent geometric or roadside improvements that came to be expected in federal-aid RRR. These problems sometimes led to misunderstandings between federal and state personnel, and to delays in project implementation. As a consequence, some states chose to shift some resurfacing projects from maintenance to construction programs and assign RRR project development responsibilities to their design divisions, which were more familiar with federal procedures and better equipped to prepare more complex designs.

Currently, resurfacing responsibilities are divided in many states—design divisions develop the plans for federal-aid RRR and sometimes for major state-funded resurfacing projects while maintenance divisions handle most spot pavement repairs, light resurfacing, and other pavement treatments that the FHWA continues to view as maintenance. Central office and district maintenance departments are also responsible in many states for developing the budget and program for both state-funded and federal-aid resurfacing and for conducting pavement monitoring programs that guide project selection.

Safety Program Responsibilities

All states collect records of accidents on the state highway system as a requirement for federal-aid eligibility. The data are used (to varying degrees among the states) in RRR and reconstruction project design and to identify high-hazard locations for corrective treatment. Also, all states have a special category of safety improvement projects in their construction programs. These projects are usually funded by federal-aid categorical hazard elimination and railroad grade crossing apportionments, but are sometimes funded (e.g., California, Michigan, Ohio, Washington) with additional federal or state funds. Maintenance funds are also used for minor safety improvements in all states.

Typically, collection and analysis of accident data are the responsibility of an office in the traffic operations division. In some state highway agencies the functions of this office may be limited to data processing and reporting accident statistics to the design division and the districts (e.g., Arizona, Florida). In others the safety office may take a leading role in developing projects and assembling the statewide safety improvement program (e.g., California, Michigan), a function analogous to the maintenance division's role in the resurfacing program.

In most states, district traffic or maintenance engineers receive reports of high-accident locations from the accident data system and other sources and are expected to investigate the sites and recommend solutions to problems. These investigations lead either to traffic engineering improvements (e.g., signals, signs, pavement marking), no action (if the problem appears unrelated to road conditions), or, in a small number of cases, to capital improvement projects.

Federal, State, and Local Relationships

To oversee the federal-aid program, FHWA operates a division office in each state capital. Because these offices maintain day-to-day contact with their respective state highway agencies and have first-hand experience with state highway needs and issues, FHWA delegates considerable authority to them. FHWA division administrators approve all federal-aid projects, design standards, and design exceptions. To each division office, FHWA assigns area engineers and technical specialists (e.g., safety engineers) who monitor federal-aid projects in different parts of the state.

Under an administrative arrangement known as certification acceptance, FHWA may delegate some of its federal-aid program oversight responsibilities to the state agency. Under full certification acceptance procedures, the FHWA division office's only responsibilities for an individual project are to approve it for inclusion in the state's federal-aid program, approve exceptions to federal standards in the design, and conduct a final inspection after the project is completed. The state certifies that federal requirements have been met regarding design standards, public participation, competitive bidding, and other aspects. FHWA conducts periodic audits of the state's performance under certification acceptance.

Certification acceptance may be limited to certain phases of project review (e.g., it may cover construction but not design) or to certain categories of projects, for example, federal-aid secondary road projects (in which case the certification acceptance arrangement is often referred to as a secondary road plan), or local government federal-aid projects. Five of the case study states (Illinois, Missouri, New York, Virginia, and Washington) have certification acceptance covering design of federal-aid primary projects (other than Interstate projects, which are never covered by certification acceptance). Application for certification acceptance is at the discretion of the state. Although the procedure can reduce federal paperwork and speed up project development, many states have not adopted it because of the increased staff burden it entails.

State highway agencies are the direct recipients of federal aid, but as noted previously, they do not administer all federal-aid highways; consequently, federal funds are passed on to local governments. The state agency usually makes funds for federal secondary and urban highways available to local governments by either a formula distribution, a discretionary process, or a combination of the two. Regardless of how federal funds are passed on to local governments, state highway agencies remain accountable to the FHWA for the use of all federal-aid funds. State highway agencies act as the liaison on all communications between FHWA and local governments and approve project designs (but not design exceptions) before submitting them to FHWA (or on behalf of FHWA under certification acceptance agreements). In addition, highway agencies sometimes let construction contracts and supervise construction for local government federal-aid projects—services that smaller jurisdictions are often not equipped to handle.

RRR Project Characteristics

To provide an indication of the relative scale of RRR work in state highway programs, differences in spending patterns for federal-aid and non-federal-aid projects, and general types of RRR work undertaken, highway project lists obtained from the case study states were used to classify highway improvement projects into seven categories: new construction and reconstruction, resurfacing and minor widening (including pavement rehabilitation short of full reconstruction), bridge work, intersection improvements, safety improvements, thin overlays and seal coats, and other (Appendix B, Tables B-3 and B-4).

1. *New construction and reconstruction:* construction of a new road; relocating an existing route on a new alignment; major widening (i.e., adding lanes) on an existing road; or reconstruction of an existing route on approximately the old alignment where the old pavement structure is removed and replaced; in general, any project intended primarily to increase the traffic-carrying capacity of the highway system. Occasionally a state may undertake a full pavement reconstruction project, primarily because of failure of the old pavement structure, without any substantial realignment or capacity increase, but such projects are rare.

2. *Resurfacing and minor widening:* a project to overlay an existing pavement with new material, usually asphalt or concrete, when the overlay is the principal activity of the project [but excluding thin overlays and seal coats (see definition of thin overlays and seal coats)]; recycling existing surface material and reapplying the surface; substantial shoulder improvements such as surfacing or regrading; and resurfacing in combination with widening to increase lane or paved shoulder width (but excluding widening to add a new

lane). This category would also include the restoration and rehabilitation categories of RRR work. The states appear to use these terms synonymously with resurfacing as defined here, and also to describe major concrete pavement repairs. Projects may include incidental improvements related to safety or traffic operations. Pavement preservation and improved ride quality are the primary objectives of all projects in this category.

3. *Bridge work:* a project in which the principal activity is building a new or replacement bridge, bridge rehabilitation (e.g., widening, deck replacement or major repairs for structural soundness); minor construction improvements such as installing railings and deck rehabilitation (structural repairs to a bridge deck short of replacement). Capacity increase, safety, or structural preservation may be the primary objective of these projects.

4. *Intersection improvements:* any construction project in which the primary activity is improvement of the operation of an intersection. Includes projects ranging from installation of traffic signals to construction of new grade-separated interchanges. Other common activities are addition of turning lanes, channelization, and realignment of intersections. The objectives of these projects are to improve capacity or safety.

5. *Safety improvements:* any project in which the principal objective is to correct a specific problem and improve safety. These projects may involve the same types of work as the preceding categories, for example, installing signals at an intersection, upgrading bridge railings, resurfacing to improve skid resistance, or reconstructing a short road segment to straighten a dangerous curve. Other common projects are guardrail installation and removal of roadside obstacles. A project is placed in this category if the state identifies it as motivated primarily by safety, rather than by capacity or preservation concerns. In all of the case study states these projects are selected through a special safety improvement programming procedure, usually involving analysis of accident statistics to identify hazardous locations on state roads and programs to install a specific category of safety improvement (e.g., standard bridge railings) throughout the state highway system. Because of differences in the ways in which the states reported information, in some instances, a project that could be defined as a safety improvement may have been classified under the functional type of work [e.g., a bridge improvement intended to correct an identified hazardous condition may be classified as bridge work rather than a safety improvement (Appendix B, Tables B-3 and B-4)], but this misclassification is not frequent enough to substantially affect the percentage distributions of spending given in the tables.

6. *Thin overlays and seal coats:* any pavement overlay not substantial enough to meet the FHWA pavement treatment criterion for work qualifying for federal aid (FHWA requires federal-aid overlays to be more than $3/4$ in. thick). Excludes thin overlays that are a stage of a larger pavement project,

patching, and overlays on short road segments. These projects have the same general objectives as the heavier resurfacing projects classified in the resurfacing and minor widening category—retarding structural deterioration of the road and improving ride quality—and therefore complement the states' major resurfacing activities. However, they have shorter lifespans than heavier overlays, and many states regard them as unsuitable on high-volume roads except as a stop-gap measure. States that have systematic pavement monitoring programs often select thin overlay and seal coat projects in the same programming procedure that they select heavier overlays (e.g., Arizona, California, Washington). The states may administer and finance these projects either as capital improvements or as part of the maintenance department's operating budget. The work is most commonly performed by contract, but may be done by state personnel. Projects usually involve no improvement beyond the pavement treatment.

7. *Other:* any projects not included in the preceding categories. Typical projects are rest area improvements, lighting, painting, and building construction.

FHWA defines RRR to include all "resurfacing and minor widening" projects. In addition, bridge rehabilitation (structural repairs or improvements to a bridge deck other than replacing the deck), a part of the bridge work category, and some projects in the safety improvement category (e.g., resurfacing to improve skid resistance, or shoulder improvements for safety purposes), also fall under the federal RRR definition *(4)*.

The purpose of the FHWA RRR definition is to identify the types of work eligible for funding from the states' federal-aid primary, secondary, and urban apportionments but ineligible before the 1976 congressional authorization of RRR as a legitimate federal-aid improvement. All projects in the construction and reconstruction category have been eligible for federal aid since the inception of the federal highway program and are beyond the scope of RRR as defined by FHWA. Nearly all projects in the intersection improvements category are eligible for federal-aid primary, secondary, or urban funding (regular federal aid) as reconstruction or as traffic operations improvements and therefore are excluded from the federal RRR definition. Thin overlays and seal coats, although they have objectives similar to RRR project objectives, are excluded from the federal definition and from federal-aid eligibility.

Nationwide, about one-fifth of the states' total non-Interstate capital expenditures is for RRR projects (Table 2-1). The majority of the states' federal-aid RRR projects are in the resurfacing and minor widening category. In several of the case study states some safety improvement projects (e.g., Ohio and Michigan) and bridge rehabilitation projects (e.g., New York, South Dakota) are approved for federal-aid primary funding as RRR projects. However,

TABLE 2-1 State Highway Agency Capital Outlays on Non-Interstate Federal-Aid Projects, by Type of Improvement and Highway Functional Class, 1985

Highway Functional Class	Improvement Type														
	Right-of-Way Engineering		New Construction		Reconstruction		Major Widening		RRR		Bridge Work		Safety/Other		Total
	$ millions	%	$ millions	%	$ millions	%	$ millions	%	$ millions	%	$ millions	%	$ millions	%	$ millions	%
Rural																
Arterials	537	16.2	549	16.5	514	15.5	208	6.3	813	24.5	585	17.6	114	3.4	3,320	100.0
Collectors	136	13.6	105	10.5	132	13.1	30	3.0	294	29.3	264	26.3	41	4.2	1,002	100.0
Urban																
Arterials	733	21.1	697	20.1	449	12.9	214	6.2	505	14.6	550	15.9	322	9.2	3,470	100.0
Collectors	45	14.8	43	14.1	55	18.1	30	9.9	52	17.1	47	15.5	32	10.5	304	100.0
Total	1,451	17.9	1,394	17.2	1,150	14.2	482	6.0	1,664	20.6	1,446	17.8	509	6.3	8,096	100.0

SOURCE: *Highway Statistics 1985*, Table SF-12A.

nationwide, federal-aid projects whose main objective is safety improvement are usually funded by the federal categorical safety programs, and projects for which bridge work is the main activity are usually funded by the federal categorical bridge replacement and rehabilitation program.

On most federal-aid RRR projects resurfacing for pavement preservation is the principal activity, but many of these projects involve improvements to bridges and intersections within the project limits, roadside safety improvements, and even reconstruction of short segments to improve alignment or replace failed pavement. The extent of these ancillary improvements—lane and bridge widening, spot realignments, and roadside safety upgrading—that should be required in the course of a federal-aid resurfacing project, is the major issue in setting RRR geometric design standards or requirements for safety enhancements.

Characteristics of Federal-Aid Resurfacing and Minor Widening Projects

The typical federal-aid project in this category involves pavement resurfacing or rehabilitation, often with minor cross-section or roadside improvements within existing rights-of-way. Lanes and shoulders may be widened, shoulder construction upgraded (e.g., replacing a gravel shoulder with asphalt), selected roadside obstacles (e.g. culvert headwalls) removed or shielded, and obsolete guardrails removed or replaced. Changes to vertical and horizontal alignment are infrequent. Narrow bridges are occasionally widened or replaced as part of these projects, but the more common treatment is to install approach guardrail and upgrade existing bridge rails.

Both FHWA RRR field reviews *(2)* and the state case studies indicate that federal-aid resurfacing projects almost always include one or more geometric or roadside improvements, such as minor widening or removal of roadside obstacles, that enhance safety. Exceptions are most likely gentle topography where roadside hazards are infrequent and existing road geometry meets applicable design standards. However, the extent and nature of safety enhancements is not consistent from state to state, particularly roadside improvements beyond the shoulder edge. For example, policies vary with respect to upgrading guardrail, moving versus protecting culvert headwalls, and relocating utility poles. The case studies revealed that identical situations on RRR projects are treated differently in different states.

In addition, resurfacing and other pavement repair projects do not always take advantage of relatively simple, low-cost opportunities to improve safety. After examining completed RRR projects in 19 states, FHWA engineers reported a number of missed opportunities for such improvements even though safety had been enhanced to some degree on all projects. Opportunities were missed because low-cost and relatively simple safety improvements were not considered during design. These missed opportunities principally involved roadside improvements and safety hardware. Simple improvements often overlooked by the states include

• Improving roadside traversability through slope flattening, ditch regrading and relocation, or removal of unnecessary guardrail;
• Removing or shielding roadside obstacles such as sign supports, utility poles, and rigid mailbox supports;
• Upgrading obsolete guardrail and bridge rail systems;
• Improving traffic operations by cutting trees and brush to restore adequate sight distance and installing new or upgraded signing and pavement markings *(1)*.

Most of the projects examined by FHWA engineers were designed before Congress enacted the requirement that all federal-aid RRR projects enhance highway safety. Review of the 15 case study states indicated that low-cost safety improvements are receiving greater attention in RRR project design as a result of FHWA rules implementing the new congressional provision. Most officials in state highway agencies and FHWA division offices indicated that roadside safety improvements are more frequently included in RRR projects now than before 1983. However, substantial variability still exists in state practice.

This variability appears related in part to differences among the states in typical existing geometric and roadside conditions (frequent occurrence of roadside obstacles may result in protection or signing rather than more costly removal) and traffic volumes (at very low volumes, higher cost roadside improvements are less likely). However, not all of the variability can be explained by these factors; it also arises from

• The practice in some state highway agencies of not using federal-aid on RRR projects when its use would require improvements that they would regard as unreasonably expensive;
• The severe restriction in some states on the funds available for pavement preservation, which necessitates stretching resurfacing dollars as far as possible;
• Differences in the stringency of FHWA division office policies regarding requirements for RRR project design and procedures for project review because of minimal guidance (before 1983) from FHWA headquarters on RRR design policy; and

- Differences in general philosophy among the state highway agencies regarding the priority and value of safety improvements, reflected in the states' programming decisions and design procedures.

In the sections on RRR project programming, finance, and design, the influence of the preceding factors on the outcomes of state RRR activities will be described.

Characteristics of Special Safety Improvement Projects

The primary intent of special safety improvement projects is to correct specific hazards. These projects are usually selected through a special programming procedure that involves screening state accident records to identify high-hazard locations and also targeting specific categories of potential hazards (e.g., substandard guardrail) for statewide treatment. They are usually funded by the federal-aid hazard elimination and railroad grade crossing programs, but additional federal or state funds are used in some states. The states usually do not regard these projects as RRR, but they are related to RRR because they involve the same types of safety improvements that many states' RRR standards require in the course of federal-aid resurfacing projects.

Safety improvement projects are generally spot treatments such as installation of new guardrail and traffic control devices, pavement marking, and skid-resistant pavement treatments. Occasionally, more costly improvements, such as intersection reconstruction or straightening a hazardous curve are made.

Project Programming

In directing their highway programs, state agencies must balance a number of competing objectives, the principal ones being the preservation of roads, improved service levels, and enhancement of safety. Success in meeting these objectives depends on the quality of individual project designs and project programming decisions—the allocation of resources to categories of construction and maintenance activities and the selection and scheduling of projects. Although the states rely heavily on federal aid for financing construction, they retain the major responsibility for programming, with the federal role limited to setting funding levels for broad program categories and establishing standards and other requirements governing design and construction of individual federal-aid projects.

Thus, while the federal government may set RRR design standards intended to encourage preserving existing highways and enhancing highway safety, the extent to which these objectives are actually met in a state's highway program as a whole depends on how they are considered in the state's programming decisions. As described in this section, the state case studies revealed that RRR programming decisions are influenced by federal RRR standards, and an understanding of these influences is necessary to evaluate the effect of RRR standards on safety and preservation.

Determining Overall Program Objectives

Although the professional staffs of highway agencies have latitude to select the construction projects that, in their judgment, best further the objectives of the state highway program, they must also comply with many restrictions, imposed by the state legislatures and the federal government, on how funds may be spent. For instance, legislative designations of state funds for particular uses, or legislative requirements for geographic allocations of funds, frequently limit the programming flexibility exercised by the agency's professional staff. Similarly, categorical spending apportionments for federal aid constrain state programming decisions. Therefore, in practice, state highway agency programming decisions reflect a compromise among the judgments of agency professionals, the state legislature, and the federal government as to how best to meet the priorities of the public at large.

Review of state highway agencies for this study generally indicated that public perceptions regarding highway needs, as evidenced by letters and conversations, public hearings, and communications with elected officials, have focused on preserving the road system (e.g., pothole repair), capacity and service-level deficiencies (e.g., congestion), and special local concerns (e.g., access to new development). Although these concerns vary from state to state, system preservation has consistently been cited as a high priority public concern. This concern has been mirrored by state legislatures that have recently increased state motor fuel taxes to fund infrastructure preservation, or have enacted statutes that set spending minimums for preservation work (e.g., Michigan, Mississippi).

Safety is also a public concern, but state officials reported that often the public's perception of safety needs does not coincide with objective indicators of safety problems, and public demands tend to focus on a few highly visible problems rather than on the types of modest, low-visibility improvements (e.g., roadside obstacle treatments and lane widening) that enhance system-wide safety.

State highway agencies reflect local public and legislative concerns in setting priorities for their construction and maintenance programs. The concern for preservation is most clearly reflected in the project programming decisions of the highway agencies in the case studies. Preservation-oriented work, from seal coats to reconstruction, is rapidly becoming the dominant component in most states' capital improvement programs. State highway agencies attempt as best they can to adapt the federal-aid program, with its various restrictions on how funds may be used, to match their own objectives. Federal aid for RRR is no exception; in spite of the program's dual-stated goals of preservation and safety, states have used federal RRR funds primarily to help achieve their road preservation objectives.

RRR Programming Procedures

As noted earlier, resurfacing or pavement rehabilitation is the primary activity of most federal-aid RRR projects, which may also include minor geometric and roadside improvements. In all 15 case studies, highway agencies selected projects of this type primarily on the basis of pavement condition regardless of funding source. Safety needs play an insignificant role in RRR project selection, and are first considered during the design process, after project selections have been made. Only occasionally when federal-aid primary or secondary funds are used for safety improvement projects such as guardrail replacement and pavement marking is safety a key factor in selecting a project that qualifies as federal-aid RRR work—but only for projects intended primarily as safety improvements, not for resurfacing and minor widening.

Although all of the state highway agencies in the 15 case study states considered pavement condition, their approaches to measuring pavement needs and selecting specific projects varied widely (Appendix B, Table B-5). Six of the states use mechanical means to measure systemwide pavement roughness, deflection, or skid resistance; eight rely on systemwide visual surveys of pavement conditions; and six do not use any type of systemwide pavement survey. In seven states the central office has the primary responsibility for selecting RRR projects, district or regional offices have this responsibility in four states, and the responsibility is divided in four states. Although there is a trend toward more systematic and objective methods of selecting resurfacing projects, professional judgment and knowledge of local conditions continue to be major factors in project selection and are unquestionably the primary means of determining the type of surface treatment or repair.

Choosing Between Federal and State Funding for RRR

An important step in developing a state's highway construction program is assigning a funding source—one or more of the federal-aid program categories or full state funding—to each scheduled project. In this procedure the state must ensure that none of its apportioned federal funds are lost because of failure to use them in the required time period. At the same time the state attempts to minimize the extent to which the restrictions on the various federal-aid categories—regarding the kinds of projects eligible and standards that must be applied—cause it to deviate from carrying out the projects it judges to have highest priority. Marked differences were observed in the criteria the states have adopted in choosing between state funding and federal aid for the RRR projects in their programs. The important factors the states consider in making this decision are

* *Availability of state money for resurfacing:* Some states rely so heavily on federal aid for all resurfacing that their actual choice for most projects is either to fund RRR work with federal aid or to defer it.
* *Cost of complying with federal RRR standards:* Required geometric or roadside improvements can be very costly on substandard roads, or may be regarded as infeasible on roads with narrow right-of-way and extensive adjacent development, but on roads built to high standards the costs of federal RRR requirements are often small.
* *Benefits of meeting federal RRR standards:* States are often more willing to undertake expensive geometric or roadside improvements on high-volume arterials and roads with poor safety records than on roads that have no history of operational problems.
* *Size (total cost) of the project:* Because of the fixed costs of federal paperwork and review procedures, states prefer to fund large projects with federal aid and small ones with state funds.
* *Urgency of the project:* Several states reported that the time required from identification of a resurfacing need to completion of the work can be less than a year if the project is state funded rather than federally funded.

All the case study states consider the second factor, cost of complying with federal RRR standards, to some extent in choosing a funding source for a RRR project, but the weight placed on this factor varies considerably. Most of the 15 case study states follow, in general, one of the following three models in choosing funding for resurfacing:

* The state has substantial state funds available for resurfacing and selects for federal aid those projects that can meet applicable RRR design standards without costly geometric or roadside improvements (Arizona, Florida, Ohio, Virginia). This strategy maximizes the miles of pavement repair within the total federal and state resurfacing budget. In developing its federal-aid RRR program, the state agency employs some means of informally screening

projects, comparing existing conditions to RRR standards to identify projects already in or close to compliance. State funds are used for projects for which compliance with federal standards would be costly. The FHWA division office often concurs in this practice or even assists in the screening to avoid federal or state conflicts over project design.

- State funds for resurfacing are limited, thus only a few state-funded RRR projects may be undertaken per year, or the only state supplement to federal-aid RRR is a small maintenance resurfacing program suitable only for stop-gap measures or low-volume roads (California, Michigan). A state in this category can be much less selective in choosing which RRR projects receive federal aid because the only alternative for most projects is to defer adequate pavement repair indefinitely. These states still attempt to stretch their federal-aid resurfacing expenditures as far as possible by deferring those projects for which compliance with standards would be the most costly and pavement repair needs are less than critical. In exceptional cases in which pavement repair cannot be delayed and the state is unable to reach agreement with the FHWA regarding geometric and roadside upgrades it finds acceptable, full state funding may be found for a major RRR project.

- Substantial state funds are available for resurfacing, but the cost of compliance with federal standards is not the predominant consideration in choosing between federal and state funding for resurfacing in Mississippi, New Jersey, and Washington. These states explicitly consider not only pavement preservation, but also the moderate capacity increases and service enhancements obtained from minor widening, spot realignment, or intersection improvements in projects constructed to RRR standards in selecting resurfacing projects to fund with federal aid. They therefore tend to select for federal aid the more complex projects on high-volume roads that have as objectives service improvement, as well as pavement repair, and use state funds for simpler projects on roads already in compliance with standards or on low-volume roads where the benefits of full RRR treatment would be small. Thus, states in this category also screen resurfacing projects for federal-aid suitability, but the objective of the screening is not simply to minimize the nonresurfacing costs of federal-aid RRR. Federal-aid RRR is only a small fraction of total annual mileage resurfaced in most of these states.

Although the strategies differ, in every case study state the federal requirements for RRR geometric or roadside improvements affect project selection. This influence may work against obtaining the best combination of system-wide safety improvement and road preservation with the funds available. For example, in several case study states, fully state-funded resurfacing projects involved pavement work only, regardless of geometric or roadside conditions (although the design of state-funded projects approached federal standards in other states). In states dependent on federal aid for RRR, resurfacing of some roads may be deferred such that pavement cost are higher than they otherwise would have been had the project not been delayed.

Financing RRR Work

Although resurfacing and minor widening projects are the predominant types of federal-aid RRR conducted in the case study states, they do not dominate total federal-aid spending on the states' federal-aid primary, secondary, and urban systems. In 7 of the 15 states these projects accounted for less than 20 percent of federal-aid expenditures in the year for which the data were tabulated, and in all states except Ohio, South Dakota, and Washington the percentage is less than 40 percent (Appendix B, Table B-3). Evidently the shift in the states' priorities from new construction to preservation does not necessarily imply that RRR has become the major federal-aid work category. Much of the new preservation work is reconstruction that entails significant capacity increases and bridge replacement or improvement projects funded largely through the federal-aid categorical bridge program.

Many states continue to rely on their own funds rather than federal aid for a substantial share of their resurfacing activities (Appendix B, Table B-6). Eight of the 15 case study states spend more on state-funded resurfacing than on federal-aid resurfacing; of the 12 states for which resurfacing mileage data were obtained, 10 resurface more miles using state funds than federal funds. The portion of federal-aid system miles receiving federal-aid resurfacing in a year was less than 2 percent in at least nine case study states. Mileage comparisons somewhat understate the importance of federal-aid resurfacing because on state-funded projects the tendency is to use thinner overlays on lower volume roads, but it is clear that most states have some flexibility in choosing between federal-aid RRR and state-funded work to address their pavement preservation problems.

Nonetheless, reliance on federal aid for resurfacing is substantial in many states and has been growing in recent years. Use of federal aid for RRR work largely follows from the increasing emphasis on preservation needs, but it also follows from shortages or constraints on state funding. As state motor fuel revenues declined and construction costs escalated during the mid- to late 1970s, some state highway agencies that had initially resisted using federal aid for RRR work felt compelled to begin doing so.

Now that state highway tax increases have been enacted in many states in recent years, some states have reduced, or plan to reduce, their use of federal aid for resurfacing (e.g., Florida and Missouri). The state of Virginia has consistently resisted using federal aid for resurfacing work, but has resorted to using it recently, not because of shortages in total capital resources but to release state funds needed to undertake urban system reconstruction projects for which federal-aid urban funds were not available.

Despite these examples, the trend toward increasing reliance on federal aid for resurfacing and other RRR work appears likely to continue. The federal RRR program is responsive to the mounting preservation concerns of state highway agencies, and increased federal-aid authorizations since the 1983 increase in federal road user taxes will tend to encourage use of federal aid for all eligible work including RRR. Increased federal aid means increased matching fund requirements so that more state funds become tied up in federal-aid work and are unavailable for full state funding of other projects.

In the early years of federal RRR funding, some states experienced difficulty in meeting the requirement that at least 20 percent of their primary and secondary federal aid be used for RRR work because they were not prepared to use federal aid for resurfacing. Apparently some FHWA division offices were willing to classify work bordering on reconstruction as RRR to help meet this requirement, and little, if any, federal aid authorization was allowed to lapse. Currently the requirement is at least 40 percent for RRR and reconstruction. None of the state highway agencies in the 15 case study states reported a problem meeting this requirement, a result consistent with GAO findings *(3)*.

Financing Special Safety Improvement Projects

In most of the highway agencies in the 15 case study states, special safety improvement projects (the primary purpose of which is to improve safety by correcting specific hazards) are funded only up to the limit of the state's federal-aid hazard elimination and railroad grade crossing apportionments. In the past a few states have accumulated large unobligated balances of hazard elimination apportionments and avoided the lapse of their federal safety funds only by developing occasional large safety projects that could absorb their safety apportionment balances.

Several states, however, (California, Michigan, Ohio, and Washington) routinely use regular federal-aid primary, secondary, or urban funds or 100 percent state construction funds to substantially supplement categorical federal aid for safety improvements (Appendix B, Table B-7).

California and Ohio have agreements with their FHWA division offices linking state spending for special safety improvements to requirements for upgrading safety in RRR projects. In California, the state highway agency has earmarked funding for a statewide priority bridge rail upgrading program in lieu of any requirement in the state's federal-aid RRR standards that bridge rail automatically be upgraded in the course of RRR work. The rationale for this arrangement is that priority selection of bridge rail improvements based on accident potential will be substantially more cost-effective than the haphazard selection that would result if bridge rail improvements were conducted solely when the bridge happened to fall within the limits of a resurfacing project. Ohio has a similar arrangement with its FHWA division office for guardrail improvements.

Design Practices and Standards

In each of the case study states, the design process for RRR projects, FHWA role in the process, design standards or guidelines used, and procedures for considering design exceptions were examined.

State Design Process

Federal-aid RRR projects are designed either in district offices or in central office design divisions. Of the 15 case study states, 9 performed most RRR design in district offices and 6 performed it in the central office (Appendix B, Table B-8). When the central office staff do not design individual projects, they usually perform review functions and are responsible for all direct contact with the FHWA regarding project approvals and design exceptions. In states with certification acceptance agreements covering design, the central office takes responsibility for the reviews and interim approvals the FHWA would otherwise perform.

Once a project is programmed, the responsible state design group prepares the plans, specifications, and estimates necessary to request contract bids. (Sometimes, final cost estimates may be prepared by a separate group within the agency.) For federal-aid RRR projects, designers and the various reviewers use a number of techniques to identify and respond to opportunities for geometric and safety improvements during this process. The techniques used vary from state to state (Appendix B, Table B-8) and include

- *Field reviews.* Project designers almost always perform predesign field reviews of existing conditions. In some states, central office staff or FHWA staff may participate in the predesign field reviews or in follow-up reviews related to specific design exception requests. Photologs are sometimes used in lieu of site visits.
- *Predesign reports.* In some state highway agencies, designers prepare predesign reports that cover items such as existing road geometry, existing

roadside features, accident history, proposed improvements, anticipated design exceptions, and expected project costs. In some states (e.g., New Jersey and South Dakota) these reports are routinely reviewed by FHWA division offices and commonly contain the justification for any requested exceptions to applicable design standards.

- *Accident data analysis.* In 13 of the 15 case study states, highway agencies routinely review accident data as part of the RRR design process. This practice probably has become universal as a result of an April 1984 FHWA directive requiring analysis of accident history for RRR projects *(5)*. Such reviews are usually intended to help identify hazardous locations or assess the reasonableness of a possible design exception. The accident analyses may be contained in a predesign report, or treated as a separate step in the process.
- *Reviews by traffic and safety specialists.* All 15 states had central traffic and safety specialists, who in most cases oversee the federal hazard elimination program. In two of the states, these specialists routinely review or critique designs for RRR projects.
- *Cost-effectiveness analyses.* Formal cost-effectiveness analyses of individual projects are rare. South Dakota and Michigan use such analyses to consider trade-offs between cost and accident reductions for removing, protecting (with guardrail), or not treating roadside obstacles. The analyses consider the location and nature of the roadside obstacle, average traffic volumes, and frequency of run-off-the-road accidents.
- *Post-construction evaluations.* Some states, such as Washington, routinely review a sample of all recent construction projects, including RRR, to evaluate operational and safety characteristics. The results of such reviews can be used to encourage better designs on future projects.

The timing of these techniques within the overall design process also varies among the states. For example, FHWA division offices may review a RRR project before it is programmed, after programming but before a formal exception request, or after a formal exception request (Figure 2-2). Schematic representations of the design process in Illinois and Ohio (Figure 2-3) illustrate the variability in the use and timing of techniques to treat safety improvement opportunities.

Although any of these techniques may promote safety-conscious design, and it is helpful to know how commonly they are used, the case studies suggest that it would be misleading to assume that the presence of any single technique, or even group of techniques, will automatically lead to effective, safety-conscious designs. Final designs are influenced not only by design and review practices but also by the standards applied, the rigor of the exception process, and the attitudes of state designers and FHWA reviewers.

58   DESIGNING SAFER ROADS

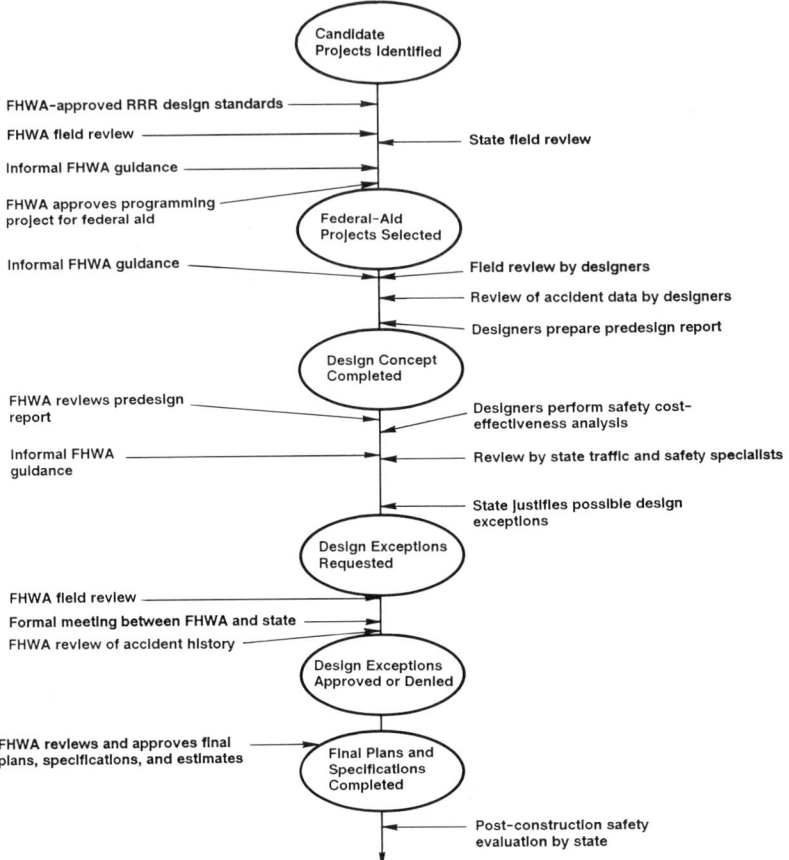

FIGURE 2-2   Generalized RRR design process (points where safety-enhancement may be considered).

Most of the highway agencies in the case study states have a two-tiered approach to RRR design—one for federal-aid RRR projects and a second for projects funded with state funds. In states where state-funded RRR projects are treated as maintenance activities, the design process may be greatly simplified, focusing almost entirely on pavement repairs. However, Illinois and New York use the same process regardless of funding, but apply less stringent standards for state-funded projects. As a result, the typical state-funded RRR project is more difficult to characterize than federally funded projects. State-funded RRR projects almost always include resurfacing, but

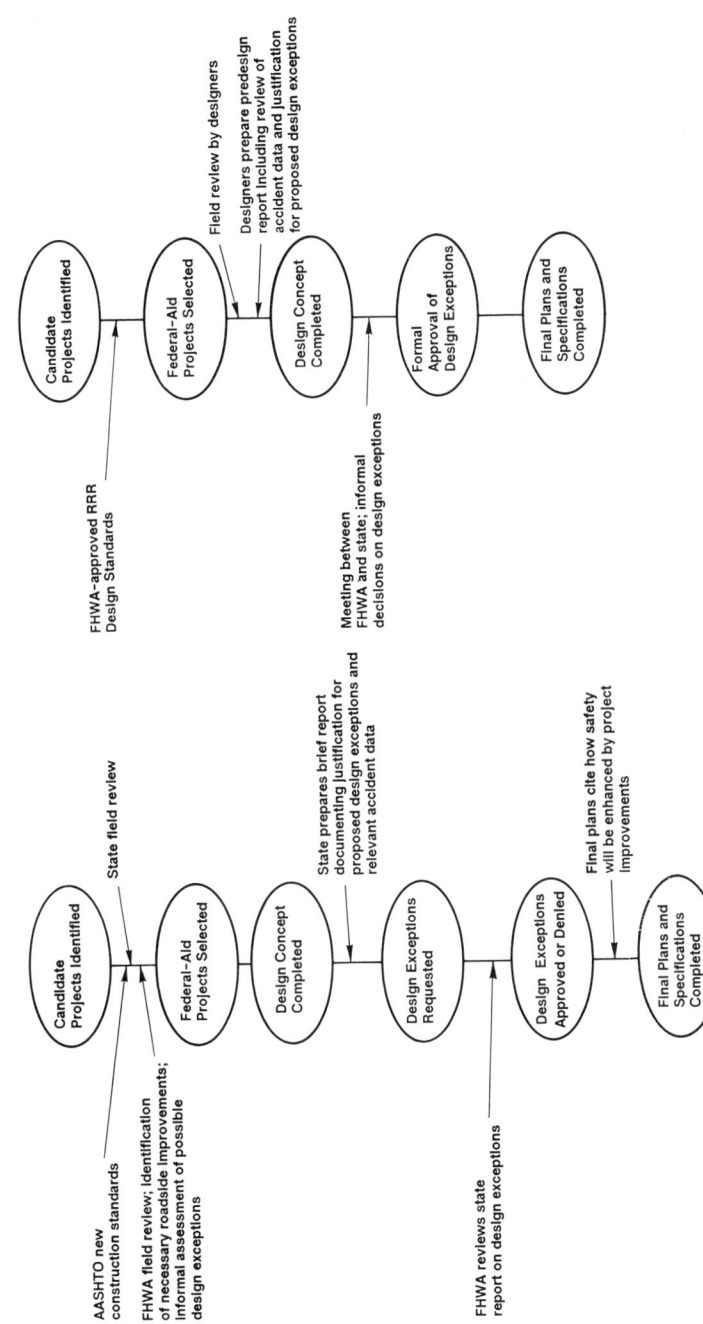

FIGURE 2-3 RRR design process in Ohio and Illinois.

geometric and roadside improvements may range from no change to improvements approaching those made on federal-aid projects.

Federal Highway Administration Role

For federal-aid construction projects, FHWA normally performs initial reviews of project scope; intermediate reviews of designs; and reviews of final plans, specifications, and estimates before authorizing construction. Responsibility for these reviews is delegated to FHWA division offices, and division administrators have the latitude to specify format and frequency of review activities. In general, the extent and detail of FHWA reviews depend on project complexity, and RRR projects are regarded as being less complex than many highway construction projects. However, an April 1984 notice to division administrators provides greater emphasis for RRR projects by suggesting that field reviews may be particularly appropriate on RRR projects ". . . to maximize opportunities to enhance highway safety" *(6)*. There are many stages at which a FHWA division office may intervene, formally or informally, in the RRR design process (Figure 2-2).

In a state where certification acceptance agreements cover design and construction, the division office's formal actions may be limited to a final design inspection. However, divisions may still perform various project and periodic program reviews in accordance with the specific certification acceptance agreements. Moreover, even when these agreements exist, the FHWA division office must approve all design exceptions on a project-by-project basis.

The role of FHWA division offices in approving design exceptions and special RRR standards varies among the states, as does the nature and extent of other project review activities. FHWA division offices in Illinois, New Jersey, and Texas have worked closely with the state highway agencies to develop special RRR standards; division offices in Ohio and New York discourage the development of special RRR standards, preferring to use new construction standards; still other division offices in Arizona and Virginia have neither encouraged nor discouraged special standards. Some division offices (e.g., New Jersey and South Dakota) require and review predesign reports for possible design exceptions; others (e.g., Texas) wait until a specific exception request is submitted. The Ohio division office conducts field reviews of all projects before programming and makes preliminary determinations about design exceptions and necessary roadside improvements for federal aid. A number of other division offices conduct occasional field reviews, often in connection with a specific exception request. Also, many division offices participate in regular meetings with highway agencies to review the status of all federal-aid projects. In some cases (e.g., Illinois) FHWA staff may make decisions regarding design exceptions at these meetings.

Most of the state highway agencies used informal contacts with FHWA to handle design issues or problems. For RRR projects, possible design exceptions are often addressed during such informal contacts, and a number of states reported that they were unlikely to pursue a formal request unless FHWA staff reacted positively to such informal inquiries.

Division office procedures and requirements for RRR projects were in transition in several states during the period in which the visits to case study states took place (spring and summer of 1984). Changes were occurring because special RRR standards had been recently proposed or approved in some states, division offices were tightening the documentation of the formal process for reviewing design exception requests [partly in response to GAO findings and recommendations (2)], and many division offices were strengthening their review of RRR projects to reflect the greater emphasis on safety required in the Surface Transportation Assistance Act of 1982. Moreover, after passage of this act, FHWA headquarters office provided more formal guidance to division offices concerning RRR policies. This guidance included memoranda to regional administrators on safety analysis (5) and design exceptions (7) for RRR projects. As a result, there is a trend toward greater federal influence in the RRR design process.

Standards

Nine of the 15 case study states had special RRR standards in effect for federal-aid projects in 1984; the remainder applied state new construction standards, which are usually based on AASHTO guidelines (8, 9). Regardless of whether or not special RRR design standards are in effect for federal-aid projects, state highway agencies and their respective division offices always have some mutual understanding concerning the geometric design values to apply, roadside improvements required, and the process for seeking and documenting design exceptions.

About one-third of the case study states had separate RRR standards for state-funded RRR projects with less stringent requirements than those applied on federal-aid projects. Other states applied the same standards for all projects but allowed more frequent design exceptions on state-funded projects. States that treat resurfacing work as maintenance did not routinely apply geometric standards or consider geometric improvements during state-funded resurfacing.

In the case study states, all special RRR standards for federal-aid projects contain minimum design values for key geometric features such as lane and shoulder widths and horizontal alignment—values generally less stringent than new construction standards (Appendix B, Table B-9). These values are compared with existing conditions at candidate RRR project sites to determine which geometric features are substandard and must be upgraded to be eligible for federal-aid. The values also may be used as a basis for selecting a particular design value when upgrading is required; however, once a state agency makes the decision to upgrade a feature to qualify for federal aid, it will in many circumstances design to new construction standards rather than to the minimum value in the RRR standards. For example, if an existing lane is 10 ft wide and the RRR standard requires 11 ft, the state often will widen to 12 ft (new construction standard) because it believes the additional cost (for a 12-ft instead of the minimum 11-ft lane) is low and consistent with a long-term goal of designing highways to full new construction standards.

The state-to-state variation of minimum values for geometric features (Appendix B, Table B-9) falls within a relatively narrow range, somewhere between AASHTO new construction guidelines and the special RRR standards proposed by FHWA in 1978. This uniformity is particularly consistent for lane and shoulder width, but less so for horizontal and vertical alignment, reflecting mutual state and federal willingness to alter cross-section characteristics more often than to undertake more costly alignment changes.

Although the range in minimum design values is narrow, the highway agencies perceive that the existing differences are necessary and accommodate unique circumstances in their states. Most officials in the case study states expressed doubts about the practicality and fairness of a single set of RRR minimum design values for nationwide application. In any event, they had largely resolved or put aside their contention with FHWA over the specific values selected for roadway geometric features and were not anxious to reopen this issue.

For roadside features beyond the shoulder edge, however, special RRR standards varied considerably (Appendix B, Table B-10); this was an issue of some contention between the state highway agencies and FHWA division offices. In some cases special RRR guidelines require the removal or protection of roadside obstacles (e.g., utility poles, headwalls, or trees), flattening of sideslopes, and replacement of outdated guardrail.

States that apply new construction standards to RRR differ in their handling of the roadside. New construction standards in New York do not address roadside obstacles beyond the shoulder edge, whereas Texas new construction standards specify roadside clear zones for all highways. Also, regardless of the new construction standards, some FHWA division offices have established guidelines for roadside treatments, even in states where special RRR standards have not been formally promulgated.

The FHWA division offices in the case study states cited a number of factors considered in establishing (or approving) design requirements for RRR projects:

- Age, design, and right-of-way of existing highways;
- Topographic and geologic conditions;
- Extent and nature of RRR project proposals;
- Extent of safety-oriented RRR activities; and
- Financial resources of the state.

Officials in Arizona, California, and New York cited increasing concerns over tort liability related to highway design. This concern appeared greatest in states that had lost sovereign immunity and had no statutory ceilings (or high ceilings) on awards. However, none of these states cited examples of tort claim concerns influencing selection of specific RRR design standards.

Because FHWA division offices differ in the extent of safety enhancements required for RRR projects, particularly on the roadside, the proportion of RRR spending on safety-related improvements varies from state to state, and RRR project designs approved for federal aid in one state may not be approved in another.

Design Exceptions

Design exceptions on federal-aid RRR projects must be reviewed and approved by the FHWA division office. The circumstances under which exceptions are required vary from state to state depending on the applicable standards for each RRR project. Usually, formal exception requests are limited to geometric features, whereas policies on roadside improvements are not regarded as standards and therefore deviating from them does not require a formal exception. However, this distinction normally is only procedural. The division offices visited simply enforced their policies on roadside improvements through interim or final approval of design plans, or other means separate from the exception review.

State highway agencies generally know what to expect before they make a formal request for design exceptions because of informal contacts with division offices and previous experience. Many state highway agencies reported that they only submit formal exception requests when they are confident of approval. Most reported that formal exception requests were usually approved, provided adequate justification was given.

A GAO review of design exceptions on RRR projects in six states revealed that vertical and horizontal alignment accounted for 44 percent of all exceptions, shoulder width accounted for 21 percent, and bridge width accounted for 13 percent *(2)*. The case studies, like the GAO review, revealed that high cost and absence of prior accidents are the most commonly cited justifications for design exception requests submitted by state highway agencies and approved by FHWA.

Procedures for evaluating exception requests differ among states, although division administrators must approve formal design exceptions. In many states, FHWA field engineers make these decisions at site reviews, meetings with state engineers, or after reviewing predesign reports or formal exception requests. Among the case study states, Texas has the most elaborate review procedure, whereby a three-member panel from the FHWA division office reviews exception requests and recommends approval or disapproval to the division administrators.

Summary

State RRR Programs

- The most common federal-aid RRR project includes pavement resurfacing or rehabilitation, often with minor cross-section or roadside improvements within existing rights-of-way. The more extensive RRR projects constitute a kind of state highway project uncommon before federal-aid RRR—one with service improvements substantially beyond the scope of simple resurfacing but with a much lower cost than full reconstruction.
- In all 15 case study states, the primary consideration for selecting potential federal-aid RRR projects is pavement condition. Safety needs and accident experience are generally not considered in project programming.
- Divergent strategies have been adopted by the states in selecting the RRR projects for which federal aid will be used. Some state highway agencies avoid using federal aid on projects that would require costly geometric and roadside improvements, seeking to maximize the miles of pavement repair per federal-aid dollar. Others tend to reserve federal aid for more complex pavement repair projects that generally include geometric improvements.
- About one-half of the state highway agencies in the United States apply special design guidelines approved by FHWA in the design of RRR projects. These guidelines are usually less stringent than new construction standards, and the design values applied by different states fall within a relatively narrow range.
- Special minimum design values for RRR projects are used to determine which geometric features on candidate projects must be upgraded to qualify for federal aid, but once the decision to improve a feature is made, many states make the improvement to the more stringent standards for new construction.
- Even in states where new construction design standards are applied to RRR projects, state highway agencies and their respective FHWA division

offices have reached informal agreements about when these geometric design values must be applied rigorously, the roadside improvements that will be required, and the circumstances under which design exceptions will be approved.

- For road features beyond the shoulder edge, special RRR standards and RRR design practices vary greatly and are of continuing contention between some state highway agencies and the FHWA.

Role of Safety in the RRR Process

- State highway agencies reflect local public and legislative concerns in setting priorities for their overall construction and maintenance programs. They report that such concerns are focused more on preservation and capacity needs than on safety, and that the emphasis on pavement repair in the use of RRR federal aid is consistent with this focus.
- The impact of the federal-aid RRR program on safety rests primarily on the extent of geometric and roadside improvements appended to a resurfacing or pavement rehabilitation project. Because FHWA division offices differ in the extent of safety enhancements, particularly on roadsides, required for RRR improvements and because existing conditions differ from state to state, the proportion of RRR spending on safety-related improvements varies from state to state. Currently, RRR project designs approved for federal aid in one state may not be approved in another. Uncertainty over the safety benefits that result from incremental design improvements underlies this variability in state and FHWA division office practices.
- The typical RRR project does include one or more improvements that enhance safety, but in recent reviews of completed federal-aid RRR projects, the FHWA has found many missed opportunities for low-cost safety improvements; most involve roadside and safety hardware. It is in the area of low-cost safety improvements, however, that FHWA division offices are evolving toward more stringent requirements.
- Nevertheless, because safety problems are not explicitly considered at the project selection stage, the impact of the federal-aid RRR program on safety is limited to modifying the design of selected projects. The federal program is unable to ensure that the selected projects will address the most urgent safety needs on a systemwide basis.

Noteworthy RRR Practices

- In general, the federally sponsored hazard elimination program is the only construction program in which state highway agencies explicitly consider safety in project selection and programming. This program is not popular with some highway agencies because it requires a substantial amount of paperwork in comparison to other programs and its funding is comparatively small. Nevertheless, some states, California and New York, for example, supplement federal hazard elimination funds with additional state funds or regular federal aid. In doing so, they can program safety-motivated projects that might not meet the federal benefit-cost guidelines for hazard elimination projects.

• In California and Ohio, FHWA division offices link the required design practices for resurfacing and minor widening projects to the states' willingness to spend regular federal aid for safety-motivated improvements elsewhere. For example, the Ohio division office requires less guardrail replacement on resurfacing projects because the state is using regular primary funds to replace its most obsolete guardrail as a separate activity.

• The mechanisms by which state agencies and FHWA division offices interact to improve the safety of specific projects vary widely and reflect unique adaptations to the specific circumstances and organizations involved.

- Special RRR design guidelines in Illinois and Texas define specific treatments for certain roadside obstacles.
- The Ohio FHWA division office conducts field reviews of prospective federal-aid resurfacing projects before programming to identify necessary roadside improvements and to make tentative decisions about design exceptions.
- Illinois and New York state highway agencies prepare predesign reports for most RRR projects, regardless of funding source, that contain a review of accident data and justifications for any proposed exceptions to applicable design standards.
- In Michigan, state traffic and safety specialists review all design plans for federal-aid RRR projects.
- South Dakota and Michigan state highway agencies perform formal cost-effectiveness analyses to determine when barrier protection is warranted for roadside hazards.
- The Washington State highway agency routinely evaluates a sample of completed RRR projects to assess safety and operational effects.

LOCAL RRR PROGRAMS

Local governments administer about one-half the road miles of the federal-aid secondary system and approximately three-fourths of the federal-aid urban system. County governments are responsible for locally administered secondary highways, and incorporated cities and towns are responsible for urban system highways. However, this division of responsibility can vary, particularly in urbanized areas where county governments may also administer some urban system highways. Further, some states allow local governments to use federal-aid bridge and hazard elimination funds. Compared with state governments, local governments are usually less reliant on federal funding. This difference occurs because local governments administer about 80 percent of the nation's highway mileage off the federal-aid system, for which federal aid is generally not provided, and because per-mile secondary and urban system federal aid is substantially lower than primary system aid.

Federal RRR requirements have less influence on local programs than on state programs because local governments are less dependent on federal aid and federal-aid highways are a smaller share of their responsibility.

RRR Project Characteristics

Local officials interviewed suggested that local governments in urban areas tend to use federal aid for major reconstruction projects whereas those in rural areas tend to use federal aid for resurfacing and minor widening projects (Appendix B, Table B-11). These tendencies are supported by data for all federal-aid construction projects that show that the dominant improvement type (excluding new construction) is reconstruction and major widening (55 percent) for urban system funds and resurfacing and minor widening (50 percent) for secondary funds (Table 2-2). On the urban system, resurfacing and minor widening account for about one-fourth of federal-aid expenditures (excluding new construction). Like state highway agencies, local governments generally view resurfacing and minor widening projects as the principal federally funded RRR activity even though other RRR work, such as bridge rehabilitation, can be funded with federal assistance.

Local officials also indicated that many urban areas currently use federal aid for intersection improvements in addition to major reconstruction projects. Such projects are not separately identified in federal project data and are usually included in the safety category (as minor traffic management improvements such as signals or left-turn lanes) or in the reconstruction category (as major intersection improvements to increase overall capacity). Intersection projects are usually not considered RRR work by either local or state agencies unless they are part of a resurfacing and minor widening project and are not covered by special RRR design standards.

Local and state officials interviewed reported that the typical federal-aid resurfacing and minor widening project undertaken by a local government is similar to a state project in this category—resurfacing or other pavement

TABLE 2-2 Comparison of Expenditures by Type of Improvement (Excluding New Construction) in Federal-Aid Urban, Secondary, and Primary Programs

Type of Improvement	Federal-Aid Program Category				
	Secondary (%)	Urban (%)	Primary (%)	Other (%)	Total (%)
Reconstruction and major widening	39.1	54.5	44.0	22.9	36.1
Resurfacing and rehabilitation Minor widening	50.3	25.7	39.9	12.6	28.0
Bridge work	10.4	17.5	15.5	62.5	34.5
Safety improvements (includes traffic control and minor intersection improvements)	0.2	2.3	0.6	2.0	1.4
Total	100.0	100.0	100.0	100.0	100.0
Total cost ($ millions)	850	959	2,409	2,944	7,161

NOTE: Compiled from FHWA Fiscal Management Information System, Form FHWA-37 (Project Status Record) data for projects completed between April 1983 and March 1984. Expenditures include state match. Expenditures for a project funded by more than one program category (e.g., a project using both federal-aid primary and other funds) are allocated accordingly among the categories.

rehabilitation, possibly with minor cross section and roadside improvements. Local officials reported that compared with a state project, however, a local project is less likely to include as many geometric improvements because the road more often meets applicable geometric standards before the improvement is made (a common situation, for example, on low-volume secondary roads, where design standards are less stringent than on high-volume roads), and cross-section design exceptions may be granted more frequently in urban areas because of environmental and community considerations.

Project Programming

Responsibility for selection of local federal-aid projects usually rests with local governments or metropolitan planning organizations that represent local governments in larger urban areas. The state highway agencies in turn review these selections, primarily to verify program eligibility, and complete some (if not all) of the necessary paperwork for securing federal funds.

Project selection and programming is more complicated in larger urban areas, where metropolitan planning organizations exist, than in small urban and rural areas. [Metropolitan planning organizations are regional organizations established in response to federal law and are responsible for planning, coordinating, and approving federal transportation investments in urbanized areas with populations greater than 50,000. Local elected officials are responsible for decision making in the metropolitan planning organization *(10)*]. Annually, the metropolitan planning organization must approve federal-aid urban projects undertaken within its jurisdiction. In some areas, constituent governments submit potential projects to the metropolitan planning organization technical staff who rank the projects using benefit-cost procedures. Projects that have large road-user benefits related to capacity increases or safety improvement (e.g., major reconstruction and intersection projects) may be assigned priority, sometimes making it difficult for resurfacing projects to be programmed with federal aid.

In other large urban areas, metropolitan planning organizations are less involved in project selection but decide how much each of their member jurisdictions will receive in annual allocations. Often informal agreements exist among members so that one jurisdiction may receive a larger share for a specific project if it is willing to forego allocations in a future year.

In smaller urban and rural areas, project selection is usually a simpler matter—the state highway agency informs the local government of its allocation of secondary or urban system aid and the local government selects the projects. In many instances the annual amounts of federal aid are small and the city or county may allow it to accrue over several years so that a major project can be undertaken.

There are exceptions to these practices. In states that maintain most secondary highways (e.g., Texas and Virginia), the highway agency retains the same key role in project selection for secondary roads that it has for other state programs. In other cases (e.g., New York) the state may not allocate its secondary and nonattributable urban system funds (urban funds not earmarked for urbanized areas with populations greater than 200,000) to local governments, instead it may make them available through a discretionary process in which local governments request project funding from the state. In such cases the state highway agency plays a more active role in the programming of local federal-aid projects.

In general, however, local agencies select federal-aid projects without a great deal of direction or pressure from state highway agencies. In the case of resurfacing and minor widening projects, local agencies often have more latitude in project selection because unlike states, they are less dependent on federal aid for this type of work. Because of this financial independence, local officials, particularly in larger jurisdictions, reported that their governments are often able to select for federal aid only those projects that do not require geometric and roadside improvements beyond those they consider worthwhile. If officials view required geometric improvements as excessive, local funds will often be used for improvements on federal-aid highways, and using local funds to resurface federal-aid highways is a common occurrence. Only in financially pressed rural counties where federal aid is a significant share of total highway funding do stringent design standards appear to cause difficulties for local officials.

Financing RRR Work

As described in the preceding section, financing responsibility is often linked to programming responsibility. When federal aid is directly allocated to local governments, they are primarily responsible for project selection; when the state makes this aid available through a discretionary process (i.e., distributing funds on a project-by-project basis), the state highway agency has some responsibility in project programming.

In either case, state highway agencies closely control the distribution of federal aid to local governments. The state highway agency most often reserves federal aid for use by a specific local government or for a specific local project; later it lets a construction contract and spends these funds on behalf of the local government. In some states, the state highway agency provides the necessary matching funds along with the federal aid it distributes to local governments. Of the highway agencies in the 15 case study states, 7 routinely provide matching funds for local urban system projects and 8 either provide matching funds or are directly responsible for local secondary projects (Appendix B, Table B-14). In cases in which matching funds are not provided, the state may allocate gasoline tax or other state revenues to local governments for highway purposes, which can be used to match federal aid.

Under federal law, urban system funds are apportioned to states based on urban population, and the portion of these funds attributable to urbanized areas with populations greater than 200,000 in each state is earmarked for use in those areas. State highway agencies may make the remaining urban system funds (about 30 percent of all urban system funds) available to local governments through a variety of means. Eleven of the case study states use population-based formulas to allocate "nonattributable" urban system funds; the remainder use a discretionary process (Appendix B, Table B-14).

Federal law formerly required that one-half of secondary funds be allocated to counties. Although this requirement no longer exists, many states continue this practice, often using the same allocation formula that the federal government uses to apportion secondary funds to the states. Eight of the state highway agencies in the 15 case study states allocate secondary funds by formula; 7 do not; of those that do not, 4 administer more than 90 percent of their state's secondary highways.

Design Practices and Standards

State highway agencies frequently share (or assume) responsibility for design and construction oversight on local federal-aid projects. Six of the state highway agencies in the case study usually assume responsibility for the design of local federal-aid projects (Appendix B, Table B-14), and all perform design reviews and coordinate with the FHWA on behalf of the local governments. Moreover, most state highway agencies generally let the construction contracts and perform construction oversight (10 of the highway agencies in the case study follow this procedure).

The federal role is less in local federal-aid projects than in state federal-aid projects. Many state highway agencies have secondary road plan agreements (14 of the case study states) or certification acceptance agreements (8 of the case study states) covering urban system projects below a certain cost, through which the state highway agency is responsible for the design reviews the FHWA would otherwise perform. Consequently, most FHWA division offices will not formally review a local secondary RRR project until the final inspection unless a design exception is required.

With regard to design standards for federal-aid RRR projects, practices for local government projects vary more than those for state projects (Appendix B, Table B-15). If a state has special RRR standards, it may or may not apply these standards for local secondary or urban system projects. Secondary projects may be covered by separate RRR standards or guidelines included in the secondary road plan agreement; new construction standards may be applied for urban projects even if special RRR standards have been developed for rural projects.

Interviews with local officials revealed mixed feelings about RRR standards. Many representatives of urban areas believed that the standards had not adversely affected their road programs for several reasons. First, major reconstruction and intersection projects were the predominant forms of work undertaken with federal aid because of local capacity and safety needs, technical ranking of projects by metropolitan planning organization staffs, or the desire to concentrate on a few large projects because of the delay inherent in using federal aid. (In the last case, local officials indicated they would rather have one or two large projects rather than five or six smaller ones delayed because of federal procedures.) Second, some urban areas could use federal aid for RRR-type work because their roads already had good design characteristics. Finally, a number of local officials indicated that design exceptions were often granted when standards presented significant problems.

However, even though design standards pose little problem, most local officials interviewed would like to have greater flexibility when using federal aid, especially given that future needs may change. Further, a significant number of local officials in both urban and rural areas indicated that RRR design standards were having negative effects on their road programs. In some cases, the standards precluded needed preservation work, forcing the locality to spend federal aid in less productive ways. In rural areas, local officials often endorsed upgrading road geometry in principle but questioned the cost-effectiveness of rigorous design standards, echoing concerns raised by many state highway agencies. In urban areas with older streets and narrow rights-of-way that cannot be widened without significant costs, some officials expressed concern that the AASHTO design standards were too stringent not only for RRR projects but also for reconstruction projects as well.

Summary

- Like state highway agency projects, the most common local federal-aid RRR project involves resurfacing with minor geometric or roadside improvements. In comparison with state programs, however, local government RRR activities are less reliant on federal aid, and as a result local governments often have greater programming flexibility. Local governments can often select for federal-aid only those projects that already meet applicable federal design standards or that would require improvements believed to be clearly cost-effective.
- In urbanized areas, geometric improvements raise environmental and community issues seldom encountered in rural areas.
- Because of special agreements under which state highway agencies assume many of the project review functions otherwise performed by FHWA for smaller projects, the federal oversight role is more limited on local federal-aid RRR projects, often consisting only of approving design exceptions and conducting final, post-construction inspections.

SUMMARY OF FINDINGS

State highway agency federal-aid RRR projects are, in general, selected and designed with the primary objective of preserving pavement and extending the life of the road. Typical federal-aid RRR projects involve pavement resurfacing and roadside improvements or minor cross-section improvements (lane or shoulder widening) within existing rights-of-way. Most non-Interstate federal-aid RRR projects are carried out on two-lane rural roads because most federal-aid road miles are of this type, and federal aid designated for urban areas is extremely limited and usually devoted to capacity improvements rather than RRR.

In the last few years, state highway agencies have paid increasing attention to safety on federal-aid RRR projects, and most projects include some improvements that enhance safety. Nevertheless, highway agencies still miss opportunities for cost-effective safety improvements on RRR projects.

Missed safety opportunities sometimes result from the standards that guide RRR project design. About one-half of the states have special standards, less rigorous than standards for new construction, that govern design of federal-aid RRR projects. The remaining states must seek approval from the FHWA to deviate from new construction standards. The special RRR standards specify the minimum values of some geometric dimensions of the roadway (e.g., lane and shoulder width) that a project qualifying for federal aid may leave unaltered, and also often identify other road conditions for which special consideration by the project designer is mandatory.

RRR standards often fail to specify firm requirements for several key design features that affect safety. In particular, the standards in use in many states set no firm requirements dictating whether the alignment of a horizontal curve, the stopping sight distance at a hill crest, the contour of sideslopes, or the width of the clear zone on a RRR project must be improved. When RRR standards do not provide definite guidance for treatment of a feature, new construction standards apply in principle, but in practice the features specifically addressed in RRR standards usually constitute the possible upgrades given serious consideration during project design. Also, new construction standards do not address all important aspects of RRR project design. For example, new construction standards do not specifically require the upgrading of obsolete guardrails.

Special RRR standards are sometimes ambiguous and subject to variable interpretation and application. For example, in many state RRR standards, minimum acceptable geometric conditions depend on the design traffic volume and design speed of the road, but the method of determining these design values is not specified. The geometric improvements dictated by the standard can differ greatly depending on whether current traffic volume or volume projected at the end of the expected life of the project is used as the design value, or whether design speed is selected on the basis of running speeds or the road's functional classification.

Standards, however, are only one factor influencing the characteristics of RRR project designs. Reviews of state highway agencies revealed many cases in which two states with similar standards produced very different RRR project designs. These discrepancies result in part from differences in the physical conditions of road systems: standards cannot provide specific guidance to fit all possible circumstances encountered on the highway system. Also, federally required RRR standards can be circumvented by using state funds for pavement repairs on roads whose existing geometry falls short of the standards.

The nature of completed RRR projects depends on the process the highway agency follows in project planning, selection, and design. This process often fails to produce a fully safety-conscious design because of a lack of (a) emphasis by top management, (b) required resources, and (c) necessary tools to identify safety improvement opportunities and evaluate options for addressing them. Little information is available about the payoffs of the kinds of safety enhancements most readily incorporated in RRR projects, and existing information usually is not accessible to designers. Project design routines place little emphasis on identifying safety needs or seeking opportunities to improve safety. The highway agency professionals most likely to appreciate the connections between geometric design and safety (e.g., traffic and safety engineers) are seldom involved in the RRR design process.

The process can also fail to produce the safest practical design because the scope of RRR projects is often too narrowly conceived, unnecessarily excluding types of improvements that could enhance safety. In many states, alignment adjustments, roadside improvements, and some types of bridge improvements are seldom considered as appropriate or feasible components of RRR designs. In most states, any improvement requiring more than a minor addition to the right-of-way is beyond the scope of a RRR project, largely because of the amount of time needed for right-of-way acquisition.

Highway agencies that produce the most safety-conscious RRR projects actively seek opportunities for safety improvements. These agencies follow systematic procedures for evaluating existing conditions to detect safety needs and analyzing a range of options to meet these needs. To evaluate existing conditions, they analyze accident records at project sites, conduct site inspections that emphasize safety opportunities and often involve safety specialists, verify existing geometric and traffic conditions for comparison with design standards or guidelines, and prepare written design reports with prescribed contents that document design procedures and safety analyses. These agencies also tend to consider a broad range of possible safety-motivated improvements as options in RRR projects, including spot alignment adjustments; attention to signs, markings, and signals; improvements at bridges and intersections; and establishment of safer roadside conditions. They look for modest, low-cost solutions in situations where upgrading to full standards would be impractical, for example, spot widening at curves and other locations prone to edge-drop problems.

Although, for the greatest impact on safety, this safety-conscious approach would extend to the planning and project selection stages of a state's highway improvement programs, serious consideration of safety in RRR generally has been limited to the design stage. In a few states, however, needs for safety improvements of the types appropriate for RRR projects are being addressed on a systemwide basis by screening the highway system for substandard geometric or roadside conditions and setting statewide priorities for their correction.

None of the case study states that produce the most safety-conscious RRR designs has comprehensively analyzed the benefits and costs of this policy. The safety emphasis usually implies higher average cost per mile for RRR projects and greater demands on the resources of the design staff. The states have not estimated the overall accident savings achieved by their policies, and these would be difficult to measure.

REFERENCES

1. *RRR Field Reviews—Final Report.* FHWA, U.S. Department of Transportation, Jan. 25, 1984.
2. *The Department of Transportation's Program to Preserve the Highways: Safety Remains an Issue.* U.S. General Accounting Office, Dec. 23, 1983.
3. *Types of Work Performed Using Resurfacing, Restoration, Rehabilitation, and Reconstruction Federal Highway Funds.* U.S. General Accounting Office, Feb. 29, 1984.
4. FHWA Notice N 5040.19, FHWA, U.S. Department of Transportation, June 28, 1976.
5. *Safety Analysis—Nonfreeway 3R Program.* Memorandum from Director, FHWA Office of Highway Safety to FHWA Regional Administrators, FHWA, U.S. Department of Transportation, April 6, 1984.
6. "Monitoring of Federal-Aid Highway Design Projects." *Federal-Aid Highway Program Manual.* FHWA, U.S. Department of Transportation, Vol. 6, Ch. 2, Sec. 1, Subsection 2, April 2, 1984.
7. *4R Program: Design Exceptions.* Memorandum from Director, FHWA Office of Engineering to FHWA Regional Administrators, U.S. Department of Transportation, July 9, 1984.
8. *Geometric Design Guides for Local Roads and Streets.* American Association of State Highway and Transportation Officials, Washington, D.C., 1981.
9. *Geometric Design Standards for Highways Other than Freeways.* American Association of State Highway Transportation Officials, Washington, D.C., 1969.
10. "Transportation Planning in Certain Urban Areas." *Code of Federal Regulations.* Title 23, Section 134.

3
Relationships Between Safety and Geometric Design

Relationships between safety and highway design features routinely improved on resurfacing, restoration, and rehabilitation (RRR) projects are described in this chapter. These relationships were used for the safety cost-effectiveness comparisons presented in Chapter 5, and can also be used by engineers making design decisions on individual RRR projects. The relationships pertain primarily to two-lane rural roads, which account for about 75 percent of all federal-aid highway mileage, 25 percent of vehicle miles traveled (VMT) throughout the United States, and 35 percent of U.S. highway fatalities. Background information on the nature and frequency of accidents on federal-aid highways is presented in Appendix H.

APPLICATION OF SAFETY RELATIONSHIPS TO DESIGN STANDARDS

The following questions about the safety effects of highway improvements were the focus of earlier debates over RRR design standards:

- What changes in accident rates can be expected if different types of geometric improvements are made?
- Will accident rates increase if highways are resurfaced without correcting existing geometric deficiencies?
- What are the safety benefits of low-cost alternatives, such as warning signs and markings, compared with more expensive geometric improvements?

Despite the widely acknowledged importance of safety in highway design, the scientific and engineering research necessary to answer these questions is quite limited, sometimes contradictory, and often insufficient to establish firm and scientifically defensible numerical relationships. Further, in those cases in which relationships can be established with substantial confidence, the results are often not known or applied by highway designers.

In general, relationships between safety and highway features are not well understood quantitatively, and the linkage between these relationships and highway design standards has been neither straightforward nor explicit. The American Association of State Highway and Transportation Officials (AASHTO), which has historically assumed primary responsibility for setting design standards, relies on committees of experienced highway designers to do this work. The committees use a participatory process that relies heavily on professional judgment. Quantitative estimates of the overall safety or cost implications of recommended design policies are not usually developed, although the process takes into account not only safety but also cost and other factors (such as the effect of design on traffic operations and capacity, maintenance implications, and design consistency for similar traffic conditions).

Although relationships between safety and highway features should be important for assessing the cost-safety trade-offs that underlie any highway standards, they are even more important for RRR standards than for new construction standards. In RRR work the costs of making incremental geometric improvements are often large relative to other project costs. For new construction and reconstruction projects, where the entire highway is being constructed from the bottom up, often on newly acquired rights-of-way, the added costs of building to higher standards are comparatively low. As a result, stringent standards (that require wider lanes, flatter curves, etc.) can be much more costly for RRR work than for new construction, and development of RRR standards requires more careful estimation of the safety payoffs expected from incremental geometric improvements.

In addition to geometric features, a variety of other factors affect highway safety, including other elements of the overall road environment (e.g., pavement condition, weather and lighting, traffic, and traffic regulations), driver characteristics (intoxication, age), and vehicle characteristics (size, weight, braking capability).

The effect of highway design is obscured by the presence of these factors. Indeed, most accidents result from a combination of factors interacting in ways that preclude determining a single accident cause. Even when a vehicle runs off the road because of driver error or equipment failure, the design of the roadside still may affect accident severity. This interaction between road, driver, and vehicle characteristics complicates attempts to estimate the accident reduction that can be expected from a particular safety improvement.

RELATIONSHIPS BETWEEN SAFETY AND KEY ROAD FEATURES

Highway features affect safety by

- Influencing the ability of the driver to maintain vehicle control and identify hazards. Significant features include lane width, alignment, sight distance, superelevation, and pavement surface characteristics;
- Influencing the number and types of opportunities that exist for conflicts between vehicles. Significant features include access control, intersection design, number of lanes, and medians;
- Affecting the consequences of an out-of-control vehicle leaving the travel lanes. Significant features include shoulder width and type, edge drop, roadside conditions, sideslopes, and guardrail; and
- Affecting the behavior and attentiveness of the driver, particularly, the choice of travel speed. Driver behavior is affected by virtually all elements of the roadway environment.

For nearly 50 years, researchers have tried to measure the effects of various road features on safety. Generally, accident rates associated with different roadway designs have been estimated by using actual accident records and travel data. The latter, usually expressed as vehicle miles of travel, is needed to express accidents relative to the number of opportunities for their occurrence.

Despite these long-term efforts, surprisingly little is known about the decrease in accident rates that result from improvements in road design. Explicit, widely accepted, quantitative relationships have not emerged. At times, researchers have been unable to agree on even the most fundamental of findings. In part, this unfortunate situation can be attributed to inherent difficulties in accident research:

- Accidents are relatively infrequent so that sound statistical studies require consistent data collected over long periods of time for many miles of highway.
- Many factors—some related to the road environment, the driver, and the vehicle—interactively contribute to the occurrence and severity of accidents. Information describing the plethora of related factors is seldom included in the accident data base. Even with reasonably complete data bases, however, researchers are often unable to sort out effects attributable to the specific roadway feature of interest. Controlled experiments are difficult to design and conduct.
- Reporting practices for nonfatal accidents differ among states and, in some cases, within states. Thus, estimates of accident rates developed using data from one area might not be appropriate elsewhere.

- Some factors, such as vehicle performance and crashworthiness, that underlie relationships between safety and road design, change over time so that relationships developed at one time may no longer be representative in later years.

Shortcomings in the way research has been organized and conducted also have hindered the development of reliable relationships between safety and road design. Knowledge about the safety effects of road design has often been uncoordinated and has lacked rigorous statistical controls. With the exception of a modest research program sponsored by the Federal Highway Administration (FHWA), few opportunities exist for coordinated, purposeful research by experienced researchers using adequate data that would over time provide the missing information about the safety effects of alternative road designs. Other countries sponsor similar research, but the results are difficult to apply in the United States because of critical differences in vehicle characteristics, traffic rules, or accident reporting practices among countries. In any event, wise investments in safety improvements are difficult to formulate in the absence of reliable safety-road design relationships acquired through a soundly managed and adequately financed research program.

With these difficulties in mind, the study committee commissioned two special research projects and several critical reviews of the existing highway safety literature in order to assess the most likely relationships between safety and the following highway design features:

- Lane and shoulder width and shoulder type,
- Roadsides and sideslopes,
- Bridge width,
- Horizontal alignment,
- Sight distance,
- Intersections,
- Pavement surface condition, and
- Pavement edge drops.

In the committee's judgment, improvements to these design features on RRR projects are most likely to have significant and measurable safety effects. Some geometric features such as cross slopes (the transverse pavement slope from the centerline on straight sections) and vertical alignment (except as a sight distance consideration) have been excluded because they do not meet these criteria. Pavement surface condition and edge drops, nongeometric features, have been included because they bear directly on the overall safety effectiveness of RRR work.

The study committee made its best judgments about the most probable relationships between safety and each of the highway design features. For each feature, the study assessed

- Whether a relationship between safety and the design features exists (e.g., is shoulder width related to safety?);
- Direction of any relationship (e.g., whether increasing shoulder width improves or degrades safety); and
- Where possible, the magnitude of the safety impact most likely over the range of improvements considered in RRR projects (e.g., the reduction in accidents expected if shoulders are widened from 2 to 4 ft).

For several of the more important features, such as lane width, horizontal curvature, and bridge width, evidence was judged to be sufficient to generalize quantitatively about the safety effects of design improvements. In the case of each feature, the generalized relationship is principally applicable to two-lane rural highways. The effects in urban settings could not be clearly documented because a substantial portion of the prior research focused on rural highways. For features such as pavement edge drops and sideslopes, development of quantitative models proved to be impossible even though considerable safety-related information was collected.

The quantitative relationships were used to estimate incremental safety benefits expected from adoption of nationwide RRR standards for the geometric design of two-lane rural roads and to assess the cost-safety trade-offs involved (Chapter 5). Also, the relationships, summarized in Appendix C, should provide guidance to highway designers who make daily decisions about trade-offs between safety and cost. Clearly much remains unknown about safety and geometric design relationships. Better understanding of these relationships should continually evolve over time as new research results become available. Current knowledge about each design feature is summarized next.

Lane and Shoulder Width and Shoulder Type

Wide lanes and shoulders provide motorists increased opportunity for safe recovery when their vehicles run off the road (an important factor in single-vehicle accidents) and increased lateral separation between overtaking and meeting vehicles (an important factor in sideswipe and head-on accidents). Additional safety benefits include reduced interruption from both emergency stopping and road maintenance activities, less wear at the lane edge, improved sight distance at critical horizontal curves, and improved roadway surface drainage.

Prior research reviewed as part of this study *(1)* indicates that

SAFETY AND GEOMETRIC DESIGN   81

- Accident rates decrease with increases in lane and shoulder width;
- In terms of accidents eliminated per foot of added width, widening lanes has a bigger payoff than widening shoulders; and
- Roads with stabilized shoulder surfaces, such as asphalt or portland cement concrete, have lower accident rates than nearly identical roads with unstabilized earth, turf, or gravel shoulders.

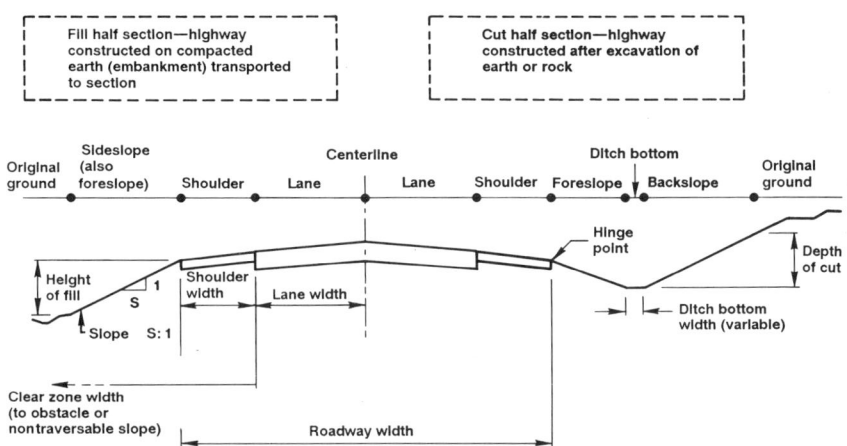

FIGURE 3-1   Cross section design features and terms.

Although the effects of individual cross-section features were estimated in several prior studies, the literature did not provide a single, internally consistent model of the simultaneous effects of lane width, shoulder width, and shoulder type on accidents. (See Figure 3-1 for an illustration of typical cross-section features.) Consequently, research was commissioned, in conjunction with FHWA, to study the combined effects of cross-section features and other variables (roadside clear zones, terrain, and traffic levels) that affect accident rates.

The research (2) produced relationships between cross-section features and accident rates consistent with the findings outlined previously (Figure 3-2). The relationships, described in Appendix C, cover single-vehicle, head-on, and sideswipe accidents. Such accidents are directly affected by lane and shoulder conditions and account for roughly 60 percent of highway fatalities on non-Interstate federal-aid highways in rural areas.

The relationships predict that widening lanes from 9 to 12 ft without shoulder improvement reduces accidents by 32 percent. Widening shoulders is less effective than widening lanes: adding a 3-ft unstabilized shoulder where

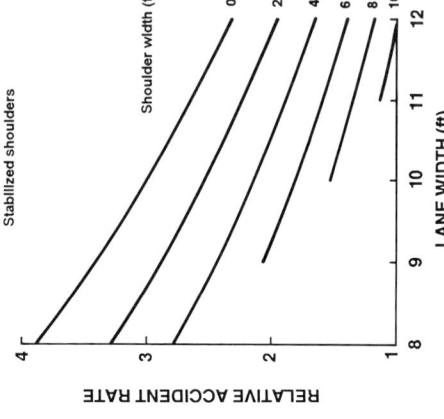
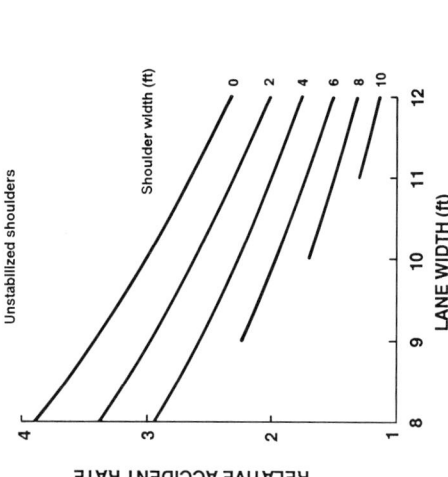

FIGURE 3-2 Normalized relationship between accidents and lane and shoulder conditions (2).

none existed reduces accidents by 19 percent. If the 3-ft shoulder addition were paved, the expected reduction would be somewhat greater—about 22 percent. The greatest gains result from a combination of improvements. For example, widening a highway with 9-ft lanes and no shoulders to 12-ft lanes and 6-ft paved shoulders reduces accidents by about 60 percent. However, the accident reduction as a result of improving a specific feature will be less when other features also are improved. For example, adding 2-ft shoulders to a highway with 11-ft lanes will eliminate fewer accidents than adding 2-ft shoulders to a highway with 9-ft lanes.

Roadsides and Sideslopes

Roadside encroachments begin when the vehicle inadvertently leaves the travel lanes, veering toward the roadside. Most encroachments are quite harmless: the driver is able to regain control of the vehicle on the shoulder and safely return to the travel lanes. When coupled with nearby roadside hazards, however, encroachments can result in roadside accidents (Figures 3-3, 3-4, and 3-5). Such accidents comprise a significant number of the accidents that occur: on two-lane rural roads, more than 30 percent of all accidents involve single vehicles running off the road *(3)*.

FIGURE 3-3 Large trees near roadway.

FIGURE 3-4 Steep sideslope.

FIGURE 3-5 Rigid drainage structure.

Past research on the safety of the roadside environment has produced important improvements to roadside hardware, including, for example, the development of barriers that better contain and more safely redirect errant vehicles and sign and luminaire supports that break away on impact, causing little damage to the striking vehicle and its occupants. In addition, design standards occasionally provide for clear recovery areas—borders beginning at the edge of the travel lanes with traversable slopes and free of hazardous obstacles. Improved designs for drainage structures such as culvert headwalls reduce the hazard posed by unforgiving roadside obstacles. Also, specifications for sideslope and ditch configuration now recognize safety benefits, as well as the more conventional objectives of construction economy, maintainability, and slope stability.

Entry of an errant vehicle onto the roadside border does not in itself mean that an accident is inevitable. Although some danger always exists, the chances of a safe recovery are excellent if the border is reasonably smooth, flat, and clear of fixed objects and other nontraversable hazards. The chances of successful recovery diminish as the ground slope within the border increases and the width decreases. Although there are no clear breakpoints, safety researchers generally agree that at speeds of approximately 55 mph, "safe" clear zones should have sideslopes no steeper than about 6:1 and should extend outward at least 30 ft from the edge of the travel lanes. When the border is flat, unintended encroachments on tangent alignments seldom extend beyond the 30-ft range.

Despite the noteworthy research described, much of what is known about roadside safety relationships remains qualitative in nature, and only tentative steps have been taken to develop comprehensive accident models. Previous studies have found significant relationships between accident rates and composite measures of roadside condition *(4-8)*. Research commissioned for this study revealed a significant relationship between the roadside recovery distance and accident rates on two-lane rural roads (Figure 3-6) *(2)*. Increasing the clear recovery area from 5 to 20 ft, for example, is estimated to reduce the number of single vehicle, head-on, and sideswipe accidents by about 35 percent.

Roadside encroachment models have been used to examine the safety effects of specific roadside features *(9-13)*. These models take into account the size and shape of a roadside feature, its distance from the travel lanes, and the probability that a collision with the roadside feature will result in an accident.

For purposes of this study, a roadside encroachment model was calibrated using the data base from a recent study of utility pole accidents (Appendix F). This new calibration was able to effectively replicate the effect of traffic volume and pole offset on utility pole accident rates. It also performed better

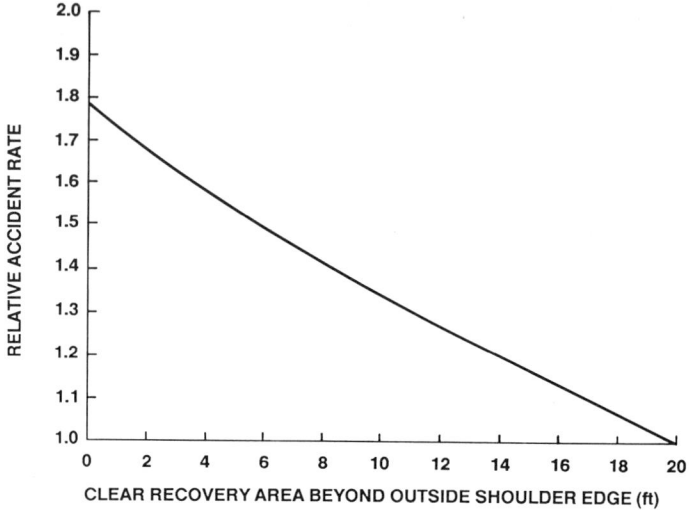

FIGURE 3-6 Normalized relationship between accidents and width of clear recovery area *(2)*.

than previously calibrated models in reproducing observed accident rates for different roadside environments. Accordingly, it is recommended for interim use in examining the safety effect of specific roadside features. Under this model, the probability of an object being struck by an errant vehicle decreases with its distance from the edge of the travel lanes such that, for example, an object located 10 ft from the travel lanes is about twice as likely to be struck as an object located 20 ft from the travel lanes.

Bridge Width

Hazards associated with bridges can be significant. Roadway constriction at narrow bridges reduces the opportunity for safe recovery by out-of-control vehicles and can result in end-of-bridge collisions. Furthermore, bridge approaches are often on a downward grade, a factor responsible for increases in speed, and, particularly in the case of older spans, are often sharply curved (Figure 3-7). When coupled with other factors such as premature icing in

FIGURE 3-7 Narrow bridge on curve.

winter and substandard bridge rail, the special hazards associated with bridges are readily understood.

Investigations of bridge safety *(14)* have revealed that bridge width is the principal factor affecting bridge safety: fewer accidents occur on wide bridges than on narrow bridges. More precisely, these studies have found the difference between the clear bridge width and the width of the approach lanes (referred to as the relative bridge width) to be a better indicator of hazard than bridge width itself (Figure 3-8). As this difference increases, observed bridge accident rates, commonly expressed in terms of total accidents per million vehicles, markedly decrease.

From several quantitative relationships developed in earlier research studies, a single one was selected for use in this study as the most likely relationship for a two-lane bridge *(15)* (Figure 3-9). This relationship, described in Appendix C, predicts that

- Increasing the difference between the width of the bridge and the width of the approach lanes from 0 to 4 ft will decrease bridge accidents by about 40 percent, with the first foot of widening accounting for nearly one-third of this reduction.
- The incremental safety gains of widening bridges decrease as bridge width increases—the first foot of bridge width beyond the travel lanes has three times the effect on accident rates as the tenth foot.

88 DESIGNING SAFER ROADS

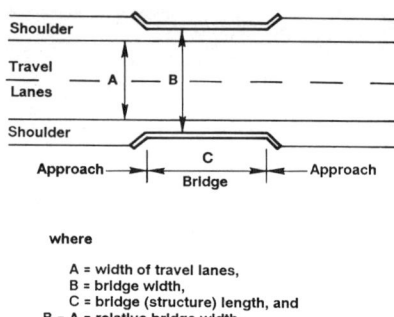

where

A = width of travel lanes,
B = bridge width,
C = bridge (structure) length, and
B − A = relative bridge width.

FIGURE 3-8 Bridge width terms and dimensions—plan view.

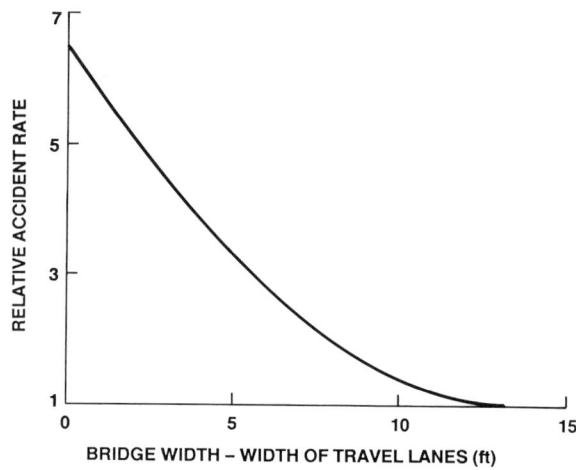

NOTE: Relative accident rate is a multiple of the number of bridge accidents per million vehicles when the bridge width minus the width of the travel lanes is 14 ft.

FIGURE 3-9 Normalized relationship between accidents and bridge width (15).

Safety at narrow bridges can also be improved by transition guardrails at bridge approaches, new or rehabilitated bridge rails, and warning signs. However, researchers have not been able to develop reliable quantitative estimates of safety benefits for these improvements.

Horizontal Alignment

Accidents are more likely to occur on horizontal curves than on straight segments of roadway because of increased demands placed on the driver and the vehicle and because of friction between tires and pavements (Figure 3-10). The safety effect of an individual curve is influenced not only by the curve's geometric characteristics, but also by the geometry of adjacent highway segments. The hazard is particularly intense when the curve is unexpected, such as when it follows a long straight approach or when it is hidden from view by a hill crest.

FIGURE 3-10 Sharp horizontal curve.

The safety effect of flattening sharp horizontal curves is of particular interest on RRR projects. When a sharp curve is improved, transitions from the straight to curved portions of the highway are smoother; the length of the curved portion of the roadway is increased; and the overall length of the highway is slightly reduced. (See Figure 3-11 for horizontal curve geometry.) Neither the combined nor individual effects of these changes on accidents are well understood.

Numerous researchers have attempted to relate changes in accident rates to specific characteristics of curve geometry, usually concentrating on degree (or radius) of curve. Past studies differ considerably in estimates of accidents per

(a) Plan view of simple curve

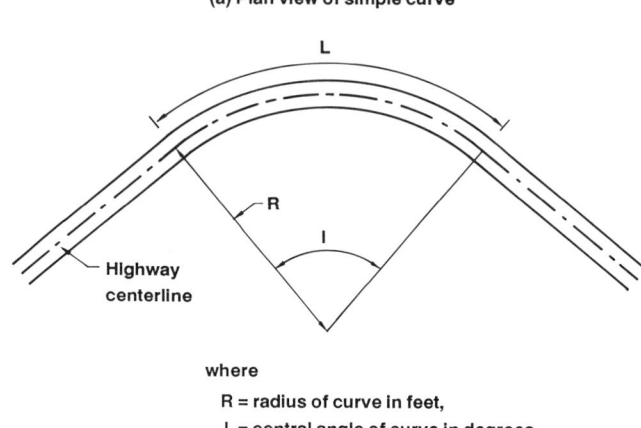

where

R = radius of curve in feet,
I = central angle of curve in degrees,
D = degree of curve = 5730/R, and
L = length of curve in feet = 100 (I/D)

(b) Cross section

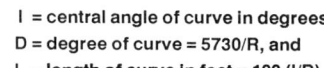

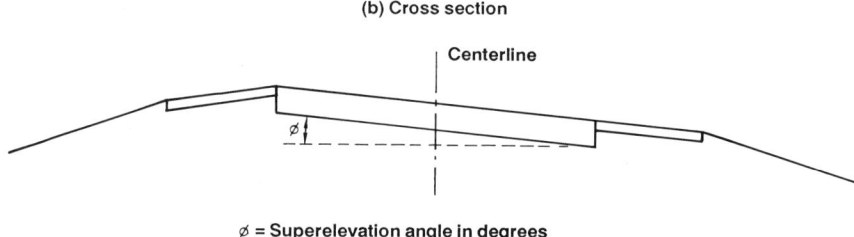

$\emptyset$ = Superelevation angle in degrees

The rate of superelevation of a roadway is the roadway slope (tan $\emptyset$).

FIGURE 3-11 Horizontal curve geometry and terms.

vehicle mile as a function of degree of curve, partly because of differences in techniques used for calculating the amount of travel and identifying accidents considered to be curve related. Also, some of the accident data bases were limited (encompassing only a single year in some cases), and influences of other geometric and traffic characteristics on curve-related accidents were not properly treated in some of the analyses (16). A recent study sponsored by the FHWA (17) succeeded in eliminating many of these problems and, in so doing, assembled the most reliable accident data base currently available for horizontal curves.

Like the data in earlier studies, the FHWA data indicate a strong link between degree of curve and accidents. The link between accidents and other measures of curve geometry, including curve length and central angle, is much weaker. For cost-effectiveness analyses of horizontal curve improvements, the study used a numerical relationship based on the FHWA data (Appendix D). In this relationship, the expected change in accidents resulting from a horizontal curve improvement is based on the change in degree of curvature, taking into account the minor reduction in road length that accompanies curve flattening (Figure 3-12).

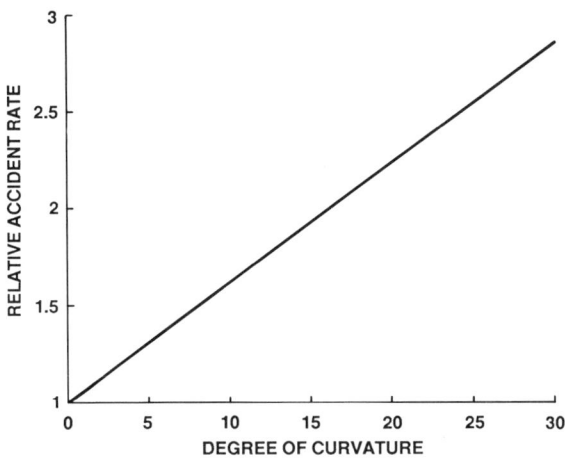

NOTES: Accident relationship is for highway segments 0.6 mi long. Relative rate is a multiple of accidents per million vehicles on tangent sections.

FIGURE 3-12 Crest curve with restricted sight distance.

The relationship between accidents and degree of curve must be regarded as rough in nature because horizontal curves are considered in isolation—without regard to the alignment of adjacent highway segments—and because the relationship does not fully correct for interrelated effects of other geometric features (e.g., sharp curves occur more frequently on roads with narrow lanes and dangerous roadsides).

As degree of curvature decreases, this relationship predicts that the number of accidents at the curve also decreases, on average by about 3 fewer accidents per degree of curvature for each 100 million vehicles passing through the curve. Flattening a sharp curve on a road carrying 2,000 vehicles per day eliminates about one accident every 8 years for each reduction in curvature of 5 degrees.

Although researchers have not been able to estimate the benefits quantitatively, a number of other design elements also affect safety at curves:

- *Adequate superelevation:* Horizontal curves on high-speed highways are usually superelevated, or banked, for safety and passenger comfort. AASHTO design policy specifies superelevation requirements based on degree of curve and design speed *(18)*. On curves where superelevation is less than that specified by AASHTO, improving superelevation as part of a RRR project is a relatively inexpensive way of increasing design speed.
- *Clear roadsides and mild sideslopes:* As noted earlier, single vehicle, run-off-road accidents are particularly common at horizontal curves. Consequently, roadsides with mild (4:1 or flatter) sideslopes would be expected to yield greater benefits in terms of reduced accident severity at curves than on tangent sections, especially on the outside of curves where more than two-thirds of the fatal, curve-related, run-off-road accidents occur *(19)*.
- *Spiral transitions:* Increased accident frequencies at horizontal curves appear to be related more to entry and exit effects than to steady-state travel on the curved roadway (Appendix D). Spiral transitions, which help drivers make smoother entries and exits, reduce the hazard at these locations.
- *Pavement surface:* Because of the vehicle dynamics involved, pavement condition on curves is particularly important for safety. Rough pavement with potholes or bumps can contribute to loss of vehicle control, and surface friction must resist lateral forces in wet weather.
- *Striping and other traffic control devices:* Striping and reflectorized markings on pavement edges and centerlines, curve warning signs, and post delineators may help drivers successfully negotiate curves at night.
- *No-passing zones:* Horizontal curves can exacerbate difficulties in carrying out passing maneuvers. The marking of no-passing zones may reduce accidents at such curves.

Sight Distance

Sight distance is the length of road ahead visible to the driver. To enhance safety on highways, designers must provide sight distances of sufficient length that drivers can avoid striking unexpected objects in the highway lanes *(18)*.

Sight distance requirements vary sharply with vehicle speeds. For example, according to AASHTO design procedures, the required stopping sight distance is 400 ft at 45 mph, 550 ft at 55 mph, and 725 ft at 65 mph *(18)*. If the increase in Interstate speed limits to 65 mph results in higher driving speeds on non-Interstate highways, sight distance improvements will become more important.

Sight-distance restrictions result from obstructions on the inside of horizontal curves, at intersections, or at sharp hill crests (Figure 3-13). Although obstructions at horizontal curves and intersections can sometimes be eliminated without changes to highway geometry (e.g., by cutting brush or trees), obstructions at hill crests can only be corrected by changes in vertical alignment—lengthening the existing vertical crest curve (see Figure 3-14).

FIGURE 3-13   Crest curve with restricted sight distance.

A recent National Cooperative Highway Research Program study for which accident data were collected for carefully matched sites with and without sight-distance restrictions due to vertical curvature, found accident frequencies to be 52 percent greater overall at sites with sight restrictions than at control sites *(20)*.

The safety effect of a sight-distance restriction is influenced not only by the sight-distance restriction itself, but also by the nature and location of any potential hazards hidden from view. For example, a heavily used but hidden intersection greatly increases the likelihood of accidents at crest curves. Without the heavy use, the necessity for rapid stopping would be greatly diminished and, as a result, so would the likelihood of stopping-related accidents. However, with sight distance problems at intersections, the hidden object is another vehicle. Thus, the 6-in. object height, which is recommended

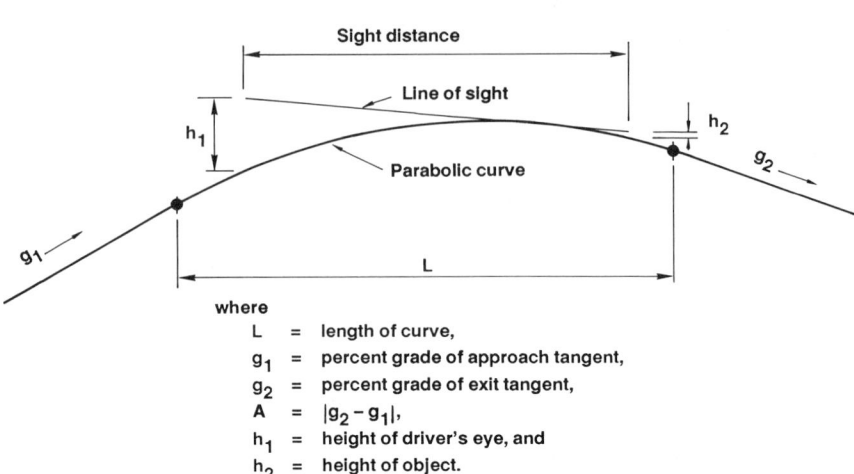

FIGURE 3-14   Vertical curve geometry, sight distance, and terms.

by AASHTO to represent small animals, rocks, or other debris in the roadway, is inappropriate for intersection problems. Rather, the object height should be the height of passenger cars (AASHTO recommends 4.25 ft in passing sight distance calculations) or, if nighttime conditions are the worst case, the height of passenger car headlamps.

The critical review of restricted-sight effects revealed no empirically based, quantitative relationship describing the influence of sight restrictions at hill crests on highway accidents (21). In the absence of such a relationship, a theoretical model, based largely on professional judgment (accident data for crest curves are not available), was used in the safety cost-effectiveness analyses discussed in Chapter 5. This model, detailed in Appendix E, can be used by highway designers investigating potential improvements at specific sites. Careful attention must be given, however, to fully understanding the type and nature of hazard at the site, and thorough analysis of available accident data is mandatory.

The model assumes that accident rates at crest curves depend on the severity of sight restriction, gradients on the approaches to the curve, and type of hazard located in the sight-restricted area. The severity of sight restriction is measured by the difference between the speed at which vehicles operate on the curve and the speed at which, according to AASHTO procedures, they could safely stop before striking an unexpected object.

As an example of expected improvement at a crest having a 9 percent gradient change and containing a "significant" hazard, as described in Appendix E, increasing the design speed from 35 to 45 mph by lowering the crest is expected to reduce the frequency of accidents about 15 percent on the 0.6-mi segment containing the curve.

Climbing lanes for slow-moving vehicles can improve safety on crest curves with inadequate sight distance for passing. Marking no-passing zones with both pavement markings and signs also improves safety at such locations.

Intersections

On two-lane rural highways, intersections are ranked together with horizontal curves and bridges as the most likely locations for accident concentration *(22)*. According to National Safety Council estimates, 56 percent of all urban accidents and 32 percent of all rural accidents occur at intersections *(23)*. Although the average accident occurring at an intersection is not as severe as one occurring on the open road, there is nonetheless a concentration of severe accidents at intersections. Of all the fatal accidents in the United States, 29 percent of those that occur on urban highways and 16 percent of those that occur on rural highways are intersection-related. It is therefore logical for safety improvement programs to place special emphasis on these natural locations of accident concentration.

Intersection improvements include changes to the physical elements of the intersecting roadways and operational measures for the control of traffic. These improvements generally focus on reducing conflict and improving driver decision making. Reducing approach speed and improving skid resistance can also be important. To achieve these objectives, intersection improvements are tailored to each individual situation, due recognition being given to traffic volumes on each of the intersecting roadways, prior accident pattern, physical characteristics of the site, and so forth. Useful procedures for selecting safety improvments at intersections include the following:

- Collision diagrams showing vehicle paths, time of occurrence, and weather conditions for individual accidents;
- Condition diagrams showing important physical features that affect traffic movement at the intersection; and
- Field review of the intersection to detect hazards not apparent from the collision and condition diagrams.

Modeling the accident effects of specific design parameters is difficult because of the large number of physical and operational features that affect highway safety at an intersection and regression-to-the-mean inaccuracies.

Also, improvements commonly address a number of intersection deficiencies simultaneously.

Although simple relationships to predict the effects of specific intersection improvements are generally unavailable, a substantial body of information exists that designers use in remedying deficiencies at hazardous sites (see Appendix G for a summary of physical and operational features affecting intersection safety). One researcher, for example, has concluded that accident reductions of 30 percent or more are possible at intersections with correctable deficiencies such as poor sight distance, inadequate signs and markings, and no channelization *(24)*.

Pavement Surface Condition

Almost all RRR projects involve resurfacing or other pavement repairs. In addition to preserving the pavement structure and improving ride quality, resurfacing also has safety implications. Indeed, substantial controversy has surrounded this point with highway organizations arguing that routine resurfacing (without geometric improvements) enhances safety and safety organizations arguing the opposite *(25)*.

The potential effect of resurfacing on safety is a result of two factors working in opposite directions. First, resurfacing reduces surface roughness and improves ride quality, generally leading to increased average speeds. Second, resurfacing often increases pavement skid resistance, which reduces stopping distance and improves vehicle controllability when the pavement surface is wet.

As part of this study, a review of available research on the safety effects of resurfacing was conducted *(26)*. This review supports the following tentative findings:

- Routine resurfacing of rural roads generally increases dry-weather accident rates by an initial amount of about 10 percent, probably because of increased speeds. Dry-weather skid resistance and stopping are unaffected by resurfacing unless the original pavement was extremely rough, so that tires did not maintain contact with the paved surface.
- Routine resurfacing of rural roads generally reduces wet-weather accident rates by an initial amount of about 15 percent. Apparently, this follows from improvements in wet-weather stopping distances and vehicle controllability that more than compensate for any effects of somewhat higher speeds following resurfacing.
- For most rural roads, the net effect of resurfacing on accident rates is small and gradually diminishes with time. Initially, the total accident rate

typically increases following resurfacing, likely by an amount less than 5 percent. When averaged over the project life, the effect of resurfacing is much less.

- Resurfacing improves the safety performance of roads that experience an abnormally high frequency of accidents in wet weather.

Resurfacing projects provide the opportunity to correct deficient pavement cross slopes at little or no extra costs. Correcting cross slopes allows better drainage of the pavement surface and improves vehicle control in wet weather. On individual resurfacing projects, careful attention to the removal of surface defects and necessary improvements to skid resistance, surface drainage, and superelevation may help offset the potentially adverse effects of increased speeds.

Pavement Edge Drops

Pavement edge drops (vertical discontinuities at the edge of the paved surface) result either from resurfacing activity unaccompanied by desirable shoulder improvement or wear or erosion of weak shoulder materials (Figure 3-15). A particularly susceptible location of edge drops is the inside of horizontal curves, owing in part to the inward off-tracking of the trailing wheels of turning vehicles.

FIGURE 3-15 Pavement edge drop.

98   DESIGNING SAFER ROADS

Detailed investigations of specific accidents have revealed the potentially deleterious effects of edge drops on vehicle safety. Although the data needed to make reliable estimates of the frequency with which edge drop problems contribute to highway accidents are not available, some safety researchers believe that vertical discontinuities, particularly at the edge of a narrow travel lane, pose a serious hazard to drivers who make otherwise minor encroachments onto the shoulder.

RRR projects provide an opportunity to correct pavement edge-drop problems, and indeed FHWA procedures require that all such problems be corrected on federal-aid RRR projects. However, resurfacing raises the elevation of the roadway unless existing pavement materials are recycled. Thus, resurfacing can increase the likelihood that edge-drop problems will reoccur, particularly where shoulders are constructed of earth, turf, or unbound gravel.

Track tests, as well as theoretical studies of vehicle dynamics, have investigated the likelihood that drivers can safely recover once they have traversed edge drops of varying heights and shapes. As vehicle speed increases, the

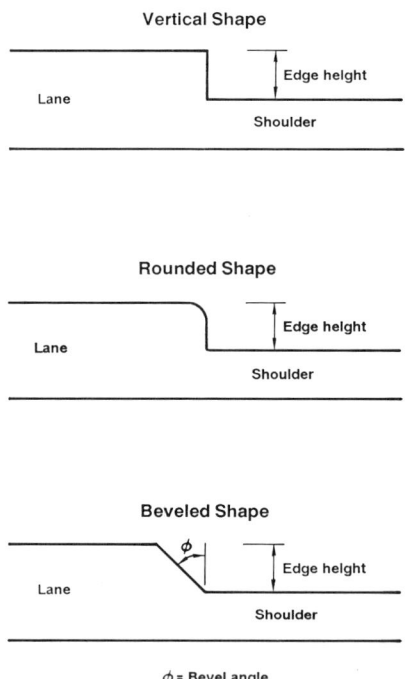

FIGURE 3-16 Pavement edge drops (cross-section views).

difficulty of successful recovery clearly increases. However, there are no generally accepted standards or guidelines on the degree of edge drop, as characterized by height and shape (Figure 3-16), that constitutes an unacceptable risk. To better understand the hazard posed by edge drops, the study commissioned additional experimental track testing *(27)* and a critical review of previous research *(28)*.

Unlike much of the earlier track testing, the new tests used ordinary as well as professional drivers, testing vertical and beveled edge-drop shapes with nominal heights of 3 and 4.5 in. for the vertical shape and 4.5 in. for the beveled shape. The tests measured the frequency with which drivers, whose vehicles had their right tires off the pavement and scrubbing against the edge, could return to a 12-ft lane without intruding into the adjacent lane. This scrubbing situation is generally the most difficult for a driver to handle.

Test results showed that ordinary drivers had more difficulty recovering from vertical drops than had been expected based on earlier tests using professional drivers. Edge drops on the order of 3 in. were generally negotiated adequately at speeds of about 30 mph in a large passenger car. However, tests on the same edge drop with small cars driven by a professional driver suggest that the safe speed would have to be lower, probably between 20 and 25 mph. Successful recovery at the 4.5-in. vertical edge was impossible at almost any speed.

Test results also showed that the shape of an edge drop affects the problems that many drivers experience in attempting to negotiate it. The use of a beveled edge, with a bevel angle of 45 degrees, was found to greatly reduce the control problems attributable to edge drops.

Both simulations and track testing require somewhat arbitrary assumptions about the nature of the inadvertent encroachments and the concept of a successful recovery, and neither addresses the frequency with which minor inadvertent encroachments can be expected. As a result, current understanding of the edge-drop hazard is incomplete. In the interim, edge drops of any height or type must be considered potentially hazardous, and should not be built into the cross section as a result of either pavement surfacing or resurfacing.

Combined Effects

Estimation of the combined effects on safety of simultaneous improvements to two or more highway features is difficult in the context of RRR work because of problems in experimental design and model calibration (Figure 3-17). The accident relationships discussed in this chapter address individual highway features and do not explicitly account for the possibility of multiple alignment changes or changes in alignment coupled with changes in cross section. However, Appendix C contains a procedure that can be used to develop reasonable estimates of the combined effect of multiple improvements.

LOW-COST SAFETY MEASURES

Highway design practice provides a broad range of low-cost safety measures that can be used to ameliorate the geometric deficiencies of existing highways. Examples include

Geometric Deficiency	*Low-Cost Safety Measure*
Narrow lanes and shoulders	• Pavement edge lines • Raised pavement markers • Post delineators
Steep sideslopes; roadside obstacles	• Roadside hazard markings • Round ditches • Guardrail
Narrow bridge	• Traffic control devices • Approach guardrail • Hazard and pavement markings
Sharp horizontal curve	• Traffic control devices • Shoulder widening • Appropriate superelevation • Gradual sideslopes • Pavement antiskid treatment • Obstacle removal or shielding
Poor sight distance at hill crest	• Traffic control devices • Fixed-hazard removal • Shoulder widening
Hazardous intersection	• Traffic control devices • Traffic signalization • Fixed lighting • Pavement antiskid treatment • Speed controls

Estimates of accident reduction factors (the percentage decrease in accidents or in accident rates expected if the measure is applied) are common for these measures. Many state highway agencies maintain lists of accident reduction factors for use in connection with hazard elimination programs, and a number of compilations from different sources have been published *(29-34)*.

State highway agencies use accident reduction factors in selecting and

SAFETY AND GEOMETRIC DESIGN   101

FIGURE 3-17   Pavement edge drop, large tree, and steep sideslope.

programming safety improvements for locations where high accident frequencies have been observed. When federal aid from the hazard elimination program is involved, the total projected accident cost savings plus any operational benefits, are expected to equal or exceed the required construction costs. Therefore, state highway agencies must estimate expected accident reductions for each federal-aid hazard elimination project.

The accuracy of the published accident reduction factors has been increasingly challenged *(35, 36)*. These challenges have arisen because most of these factors are derived from simple comparisons of accident data for periods immediately before and after a measure is applied. The comparisons often are biased toward overestimating accident reductions attributable to applied measures because the sites at which measures are applied are not selected at random (sites with recent histories of abnormally high accident frequencies are usually selected) and control sites are not used.

From the standpoint of RRR work and safety cost-effectiveness, a further problem with published accident reduction factors is that they are generally insensitive to the degree of hazard. For example, a typical factor might indicate that the installation of warning signs at curves reduces accidents by 30 percent regardless of curve geometry. Although this factor might be appropriate for very sharp curves, it will overstate the safety effects of installing warning signs at all curves with design speeds less than the speed limit.

Recognizing the need for caution in the use of published accident-reduction factors, the study committee nonetheless concluded that the low-cost safety measures described can provide significant reductions in the frequency and severity of accidents. The safety benefits of these measures, coupled with their low costs, are such that the measures can be highly cost-effective on RRR projects.

EFFECT OF CHANGING VEHICLE FLEET

To assure safe use, highways must be designed and built to accommodate the types and volumes of vehicles in operation. Because important vehicle characteristics, such as average size, weight, performance, and crashworthiness evolve over time, the compatibility between highway geometry and vehicle characteristics may be altered in ways that affect the severity and frequency of accidents. Because of these possibilities, the committee commissioned a review of new vehicle trends and forecasts and an assessment of whether fleet changes over the next 15 years are likely to cause fundamental changes in the current relationships between safety and highway design elements *(37)*. Findings of this assessment include the following:

• *Automobile and light-truck characteristics.* Changes in safety-related automobile and light-truck characteristics in the recent past were precipitated primarily by the fuel shortages and price increases of the 1970s. These forces have now diminished and are not likely to precipitate further change of significant magnitude during the next 15 years.

> *Gross vehicle weight.* Sparked by gasoline shortages and rapid price increases in the 1970s (and resulting federal fuel economy standards), U.S. manufacturers introduced substantially lighter automobiles and light-duty trucks. Between 1978 and 1980, for example, average gross vehicle weight for new automobiles dropped from approximately 3,600 to 3,200 lb. Since then, average weight first stabilized and then increased slightly as a result of size increases in the smallest imported vehicles. Because fuel prices are expected to rise only gradually and further improvements in fuel economy are possible without weight reductions, no significant change in average weight is expected through 1990 and probably through 2000.
>
> *Dimensions.* Accompanying reductions in gross vehicle weight were reductions in average vehicle length and width. No further reductions in vehicle length and width are expected through 1990 and probably through 2000.

Braking. Possibly increased use of antilock braking systems in new automobiles could ultimately reduce average stopping distances.

Driver eye height. Average driver eye height decreased as vehicles were downsized. For some vehicles, the actual height is now below the value (42 in.) assumed in AASHTO stopping sight distance standards. However, no further reduction in average driver eye height is expected through 1990 and probably through 2000.

Underclearance. Reduced underclearances have also accompanied reductions in automobile size. Underclearances as low as 4 in. are now common. This underclearance is less than the 6-in. object height assumed in AASHTO stopping sight distance policies.

- *Heavy-truck characteristics.* The history of motor freight transportation reflects a continuing increase in truck size and weight over time. Although continued upward adjustments can be expected as a result of state legislative activities, no new regulatory initiatives are anticipated at the national level during the next 5 years.

   Dimensions. The Surface Transportation Assistance Act of 1982 required states to permit longer, wider trucks on Interstate highways and major primary routes than were previously permitted by many states. Now allowed nationwide are tractor semitrailers with semitrailer lengths of at least 48 ft (usually replacing 45-ft semitrailers), twin trailer trucks with two 28-ft trailers (usually replacing the 45-ft semitrailer), and 102-in. wide trucks (replacing 96-in. wide trucks).

   Gross vehicle weight. Although the change in federal law did not generally increase truck weight limits, the increased volumetric capacity will increase average gross weights.

   Braking. New trucks are not expected to have impaired braking capability, provided that brakes are properly designed and adjusted.

   Handling. New trucks will perform differently from the trucks they replace, particularly with respect to off-tracking where the rear wheels of a turning vehicle follow a different path than its front wheels. The longer semitrailers experience more low-speed, inward off-tracking, and twins experience less inward off-tracking than the semitrailers they replace. Under high-speed conditions, twins exhibit greater outward off-tracking, and 48-ft semitrailers exhibit less outward off-tracking than the semitrailers they replace.

- *Patterns of use*

   Vehicle mix. On major highways during at least the past 20 years and possibly longer, trucks have comprised an increasing proportion of total traffic. Although this trend may well continue into the future, the rate of increase is likely to diminish.

Diffusion of large-truck operations. The Surface Transportation Assistance Act of 1982 addressed the use of longer and wider trucks on a nationwide "system" of limited mileage, but it is expected that these larger trucks will be increasingly used on other highways as well.

Growth of light trucks. Sales of light trucks (pickups, panels, vans, and utility trucks) have grown relative to automobile sales. These light trucks have a higher center of gravity than automobiles, which may have implications for barrier and guardrail designs.

Although anticipated changes in the vehicle fleet, as previously summarized, are likely to affect highway safety, the effects will be small and almost certainly mixed. For example, continued vehicular improvements, such as increased use of antilock braking systems, are expected to have positive effects on safety. At the same time, increasing disparity between the sizes and weights of automobiles and heavy trucks and the increasing proportion of heavy trucks on major highways may result in safety being degraded. These anticipated fleet changes may be sufficiently large to warrant reexamination of selected new design standards (e.g., stopping sight distance, which is influenced by both braking performance and driver eye height) and roadside hardware design (e.g., breakaway signs whose design is affected by weight of the wayward vehicle). However, the changes do not appear to be sufficiently large to measurably influence the relationships between highway safety and the roadway and roadside features affected by typical RRR improvements. Such relationships appear to be relatively insensitive to evolutionary changes in vehicle characteristics of the type experienced in the past and expected in the future.

ROADWAY CONSISTENCY

Unfortunately, safety relationships presented elsewhere in this chapter fail to capture situational influences present in the roadway environment that contribute greatly to roadway hazards *(38-40)*. Illustrative of these particular hazards are high-volume intersections in isolated rural settings, sharp horizontal curves following long segments of generally straight alignment, and compound curves—contiguous horizontal curves turning the same way—in which a flat curve precedes a much sharper one. Common to such situations is the violation of driver expectancy: the unfamiliar or inattentive driver, lulled to complacency by the gentleness of the approaching roadway, is surprised by the sudden appearance of a potential hazard. The response is uncertain and slow, leading possibly to inappropriate maneuvering and an increase in accident potential.

In addition to inconsistencies that occur at point or spot locations, such as illustrated previously, other situations are found in which clues from the physical environment belie the nature of the roadway hazard. Perhaps the most important arises from possible incompatibilities between the roadway cross section and its horizontal and vertical alignment. In the case of a roadway improvement, for example, upgrading cross-sectional elements without corresponding upgrading of alignment can result in an erroneous and potentially hazardous illusion of safety and the selection of operating speeds excessive for the critical alignment conditions.

In general, the degree of hazard inherent in a specific feature, such as a narrow bridge, sharp curve, or a roadway without shoulders, depends not only on the feature itself but also on the nature of the nearby roadway environment. Although the safety effects of such interactions have not been quantified, it is quite likely that expected safety gains, estimated by using the relationships described in this chapter, will understate achievable gains if roadway inconsistencies are eliminated as a part of the RRR improvement. Useful techniques for eliminating spot inconsistencies or compensating for their potentially adverse effects include

- Provision of gradual geometric transitions appropriate to the anticipated vehicle operating speed;
- Improvement of sight distance for early detection of the presence of the critical feature;
- Provision of gentle sideslopes with few roadside obstacles at critical locations; and
- Installation of traffic control devices appropriate for the situation.

SUMMARY

Motor vehicle safety is critically influenced by the way highways are built. Improvements to the following design features on RRR projects are most likely to have significant and measurable effects on safety:

- Lane and shoulder width and shoulder type,
- Roadsides and sideslopes,
- Bridge width,
- Horizontal alignment,
- Sight distance,
- Intersections,

- Pavement edge drops, and
- Pavement surface condition.

For the first five features listed, evidence is sufficient to support quantitative estimates of the reduction in accident rates as a result of design improvements on two-lane rural highways.

For intersections and pavement edge drops, the safety effects of design improvements are well-established, although quantitative relationships between accident rates and design improvements are not available.

Design improvements on RRR projects can affect not only accident rates but also accident severity. Past research on the safety of the roadside environment, for example, has produced roadside hardware that reduces the likelihood that an accident will result in a fatality or a serious injury. However, most of what is known about the effects of design improvements on accident severity remains qualitative in nature.

Pavement resurfacing without other highway improvements will have a small negative effect on safety on most roads. On roads with unusually high percentages of wet-weather accidents, however, resurfacing may slightly reduce the total number of accidents. In both cases, the effects on safety are small initially and decrease with pavement wear.

Although the safety effects of low-cost operational measures are frequently overstated in the literature, these measures can provide significant reductions in the frequency and severity of accidents. The safety benefits of these measures, coupled with the low cost to implement them, are such that the measures can be highly cost-effective on RRR projects.

REFERENCES

1. C. V. Zegeer and J. A. Deacon. "Effect of Lane Width, Shoulder Width, and Shoulder Type on Highway Safety: A Synthesis of Prior Literature." In TRB State-of-the-Art Report. TRB, National Research Council, Washington, D.C. (forthcoming).
2. C. V. Zegeer, J. Hummer, D. Reinfurt, L. Herf, and W. Hunter. Safety *Effects of Cross-Section Design for Two-Lane Roads—Vol. I and II*. Report FHWA-RD-87/008 and 009. FHWA, U.S. Department of Transportation, 1986.
3. J. L. Graham and D. W. Harwood. *NCHRP Report 247: Effectiveness of Clear Recovery Zones*. TRB, National Research Council, Washington, D.C., 1982, 68 pp.
4. P. H. Wright and K. K. Mak. "Single Vehicle Accident Relationships." *Traffic Engineering*, Vol. 46, No. 1, Jan. 1976, pp. 16–21.
5. D. E. Cleveland and R. Kitamura. "Macroscopic Modeling of Two-Lane Rural Roadside Accidents." In *Transportation Research Record 681*. TRB, National Research Council, Washington, D.C., 1978, pp. 53–62.

6. T. J. Foody and M. D. Long. *The Identification of Relationships Between Safety and Roadway Obstructions.* Columbus Bureau of Traffic, Ohio Department of Transportation, Jan. 1974.
7. C. V. Zegeer and M. J. Cynecki. "Determination of Cost-Effective Roadway Treatments for Utility Pole Accidents." In *Transportation Research Record 970.* TRB, National Research Council, Washington, D.C., 1984, pp. 52–64.
8. C. V. Zegeer and M. R. Parker, Jr. "Effect of Traffic and Roadway Features on Utility Pole Accidents." In *Transportation Research Record 970.* TRB, National Research Council, Washington, D.C., 1984, pp. 65–76.
9. J. C. Glennon. *NCHRP Report 148: Roadside Safety Improvement Programs on Freeways: A Cost-Effectiveness Priority Approach.* TRB, National Research Council, Washington, D.C., 1974, 64 pp.
10. J. C. Glennon and C. J. Wilton. *Effectiveness of Roadside Safety Improvements: Vol. I, A Methodology for Determining the Safety Effectiveness of Improvements on All Classes of Highways.* Report FHWA–RD–75/23. Midwest Research Institute, Kansas City, Mo., Nov. 1974.
11. J. C. Glennon. *NCHRP Report 214: Design and Traffic Control Guidelines for Low-Volume Rural Roads.* TRB, National Research Council, Washington, D.C., 1979, 41 pp.
12. *Guide for Selecting, Locating, and Designing Traffic Barriers.* American Association of State Highway and Transportation Officials, Washington, D.C., 1977.
13. T. C. Edwards et al. *NCHRP Report 77: Development of Design Criteria for Safer Luminaire Supports.* HRB, National Research Council, Washington, D.C., 1969, 82 pp.
14. K. K. Mak. "Effect of Bridge Width on Highway Safety: A Synthesis of Prior Research." In TRB State-of-the-Art Report. TRB, National Research Council, Washington, D.C. (forthcoming).
15. D. S. Turner. "Prediction of Bridge Accident Rates." *Journal of Transportation Engineering,* Vol. 110, No. 1, American Society of Civil Engineers, New York, Jan. 1984.
16. J. C. Glennon. "Effect of Alinement on Highway Safety: A Synthesis of Prior Research." In TRB State-of-the-Art Report. TRB, National Research Council, Washington, D.C. (forthcoming).
17. J. C. Glennon, T. R. Neuman, and J. E. Leisch. *Safety and Operational Considerations for Design of Rural Highway Curves.* Report FHWA–RD–86/035. FHWA, U.S. Department of Transportation, Aug. 1986.
18. *A Policy on Geometric Design of Highways and Streets.* American Association of State Highway and Transportation Officials, Washington, D.C., 1984.
19. K. Perchonak, T. A. Ranney, A. Baum, M. Stephen, and F. Dominic. *Methodology for Reducing the Hazardous Effects of Highway Features and Roadside Objects.* Report FHWA–RD–78/202. FHWA, U.S. Department of Transportation, May 1978.
20. P. L. Olson et al. *NCHRP Report 270: Parameters Affecting Stopping Sight Distance.* TRB, National Research Council, Washington, D.C., June 1984, 169 pp.

21. J. C. Glennon. "Effect of Sight Distance on Highway Safety." In TRB State-of-the-Art Report. TRB, National Research Council, Washington, D.C. (forthcoming).
22. C. P. Brinkman and S. A. Smith. "Two-Lane Rural Highway Safety." *Public Roads*, Vol. 48, No. 2, Sept. 1984, pp. 48–53.
23. *Accident Facts—1985 Edition*. National Safety Council, Chicago, Ill.
24. D. Cleveland. "Effect of Intersection Safety Improvements on Highway Accidents: A Synthesis of Prior Research." In TRB State-of-the-Art Report. TRB, National Research Council, Washington, D.C. (forthcoming).
25. U.S. Congress. House of Representatives. Subcommittee on Investigations and Oversight, House Committee on Public Works and Transportation, *Resurfacing, Restoration, Rehabilitation (3R) of Roads Other Than Freeways*. Hearings . . . 97th Congress, Sept. 17, Oct. 27, 28, Dec. 15, 1981 (Serial No. 97-75).
26. D. Cleveland. "Effect of Resurfacing on Highway Safety: A Synthesis of Prior Research." In TRB State-of-the-Art Report. TRB, National Research Council, Washington, D.C. (forthcoming).
27. P. L. Olson, R. Zimmer, and V. Pezoldt. *Pavement Edge Drop*. The University of Michigan Transportation Institute, Ann Arbor, 1986.
28. J. Glennon. "Pavement/Shoulder Drop-offs As They Affect Highway Safety." In TRB State-of-the-Art Report. TRB, National Research Council, Washington, D.C. (forthcoming).
29. J. A. Smith et al. *Identification, Quantification and Structuring of Two-Lane Rural Highway Safety Problems and Solutions*, Vol. I. Report FHWA–RD–83/022. FHWA, U.S. Department of Transportation, June 1983.
30. *Highway Safety Engineering Studies*. FHWA, U.S. Department of Transportation, 1980.
31. T. Creasey and K. R. Agent. *Development of Accident Reduction Factors*. Report UKTRP-85-6. University of Kentucky Transportation Research Program, Louisville, March 1985, 82 pp.
32. *Accident Reduction Levels Which May be Attainable From Various Safety Improvements*. FHWA, U.S. Department of Transportation, Aug. 1982.
33. Roy Jorgensen Associates. *Evaluation of Criteria for Safety Improvements on the Highway*. Gaithersburg, Md., 1966.
34. Roy Jorgensen Associates. *NCHRP Report 197: Cost and Safety Effectiveness of Highway Design Elements*. TRB, National Research Council, Washington, D.C., 1978, 46 pp.
35. E. Hauer and J. Lovell. "New Directions for Learning about the Safety Effect of Measures." In *Transportation Research Record 1075*. TRB, National Research Council, Washington, D.C. 1986, pp. 96-102.
36. K. S. Opiela. "Evaluating the Effectiveness of Highway Safety Improvements." Presented at the ASCE Nashville Conference on Highway Safety, Nashville, Tenn., March 1986.
37. W. D. Glauz. "Effect of Possible Future Change to the Vehicle Fleet on Highway Safety." In TRB State-of-the-Art Report. TRB, National Research Council, Washington, D.C. (forthcoming).
38. G. J. Alexander and H. Lunenfeld. *Positive Guidance in Traffic Control*. FHWA, U.S. Department of Transportation, April 1975.

39. T. J. Post et al. *A Users' Guide to Positive Guidance*. FHWA, U.S. Department of Transportation, June 1977.
40. C. J. Messer, J. M. Mounce, and R. Q. Brackett. *Highway Geometric Design Consistency Related to Driver Expectancy*, Vols. I and II. Report FHWA–RD–81/035 and FHWA–RD–81/036. FHWA, U.S. Department of Transportation, April 1981.

4
Relationships Between Highway Costs and Geometric Design

This chapter contains a discussion of the relationships between cost and key highway design features. More specifically reported are the added costs highway agencies incur on resurfacing, restoration, and rehabilitation (RRR) projects when existing road geometry is improved (e.g., through lane and shoulder widening or horizontal curve reconstruction). Along with the safety-geometric design relationships described in Chapter 3, these cost data are the key elements of the safety cost-effectiveness analyses presented in Chapter 5.

In the sections that follow, the inherent problems of generalizing about highway cost are reviewed, and typical RRR project costs obtained from selected states (Chapter 2) are reported. The relationships between RRR project cost and incremental improvements to existing road geometry are presented followed by discussions of the right-of-way requirements for RRR projects and the longer term maintenance implications of improved highway geometry.

COST RELATIONSHIPS—PROBLEMS AND LIMITATIONS

The scope of geometric improvements included in a RRR project affects highway agency costs in two ways. First, it can increase or decrease the initial capital investment required for the design and construction of the RRR project, including right-of-way acquisition. These costs are referred to in this report as RRR project costs. Second, changes to highway geometry may affect longer term maintenance requirements. Initial RRR project costs tend to be the dominant consideration for highway agencies making decisions about geometric design improvements. In some cases, however, maintenance costs also can be important, and neglecting to take them into account can result in poor design decisions. For example, placing a guardrail may appear to be an attractive alternative to flattening sideslopes because the initial project cost is less; however, a different conclusion may be reached when the long-term cost of maintaining the guardrail is taken into account. Also, in snowbelt states the placement of quardrail might lead to higher costs for plowing and sanding because of problems with drifting snow.

Although maintenance costs can be an important consideration in decisions about geometric design improvements for individual projects, the establishment of generalized relationships between maintenance costs and specific types of design improvements is very difficult. There are many different components of maintenance costs, some of which may be increased and others of which may be decreased by a specific type of improvement. For example, paving shoulders increases the amount of paved surface to be maintained but may reduce pavement raveling by moving the pavement edge further from traffic. Because of these difficulties, the cost relationships presented in this chapter do not include maintenance costs.

Unfortunately, a variety of factors make it difficult to relate RRR project costs to incremental geometric improvements in ways that could be used to calculate the potential cost impact of particular design standards. These factors include

- *Variable site conditions.* Costs for specific geometric improvements, such as widening lanes from 10 to 12 ft or increasing a horizontal curve radius from 500 to 1,000 ft, can vary greatly depending on site-specific factors such as topography, right-of-way availability, and drainage requirements.

- *Variable labor and material costs.* Costs for identical geometric improvements on similar sites can vary within a given state, as well as between states because of different construction unit costs. Labor costs in the San Francisco area, for example, are nearly double those in Jackson, Mississippi *(1)*. Even within a given state, cost variation can be substantial; a resurfacing or widening project in south Florida (Monroe County) may cost 40 percent more than an identical project in the Florida panhandle *(2)*.

- *Variable design practices.* Highway agencies differ in the way they design pavements, shoulders, drainage structures and other items not directly related to geometry. For example, New York State routinely paves highway shoulders, whereas Virginia constructs gravel or turf shoulders; Ohio often applies 1-in. overlays; Washington State applies 1.5- to 2.0-in. overlays. Such differences cause variations in cost even when the geometric improvements are the same, the sites are similar, and unit prices are identical.

- *Variable project scale.* Unit prices for construction can vary depending on the quantity involved so that cost estimates based on average unit cost can mask the effects of economies of scale related to geometric improvements. For example, review of sample RRR projects in Washington State revealed excavation unit costs 10 to 15 percent less for projects where more than 10,000 yd^2 of earth were excavated compared with 3 projects where between 1,000 and 10,000 yd^2 were excavated *(3)*. Explicitly incorporating economies of scale into RRR cost estimates, however, is complicated by the general lack of quantity-sensitive unit cost data and the reality that the scopes of RRR projects are often arbitrary. For example, the length of a project may be based on the limits of the last resurfacing project or the level of funding set aside. Because material quantities for resurfacing and lane and shoulder widening are related to project length, unit costs will increase or decrease depending on the length selected.

• *Complementary construction requirements.* In addition to promoting economies of scale, multiple geometric improvements on a RRR project may entail complementary construction requirements that can further reduce incremental cost. For example, lengthening a crest vertical curve to improve sight distance involves excavating earth that must be transported off-site if there are no fill requirements for other improvements. The material could, in some cases, be used for flattening sideslopes or widening shoulders.

• *Scale of historical cost data.* Highway agency cost data are often available at levels that are either too coarse or too detailed to be directly useful for estimating the effect of specific geometric improvements on RRR projects. At one extreme, costs may be available on a per-mile basis for different project categories; for example, resurfacing and widening projects on two-lane rural arterials in a given state might cost $250,000/mi. Such estimates are useful for early budgeting and programming, but shed no light on incremental costs for widening 10-ft lanes to 12 ft. At the other extreme, most state highway agencies maintain records on unit costs for hundreds of construction items such as excavation, subbase, or guardrail installation. These data are used in conjunction with quantity estimates to develop detailed project cost estimates. However, estimating the effect of specific geometric improvements on quantities for each construction item can be a difficult and time-consuming task.

• *Project costs bias.* Cost estimates for incremental geometric improvements based on typical project costs can be biased because they exclude projects rejected because of high cost. Hence, if flattening horizontal curves typically costs $150,000 each on a completed RRR project, the cost of flattening all similar curves would probably be higher.

Because of these factors, no single set of universally applicable cost relationships can be developed for incremental geometric improvements. At best cost relationships can be developed that indicate the added costs for geometric improvements on RRR projects typical of a particular state or region. Such relationships can be used in conjunction with accident relationships to examine the safety cost-effectiveness of particular geometric improvements in a state or region.

TYPICAL RRR PROJECT COSTS

As described in Chapter 2, the typical federal-aid RRR project involves pavement resurfacing, often with minor lane and shoulder widening or roadside improvements. Occasionally, spot improvements to vertical or horizontal alignment are undertaken as part of a RRR project. Similar projects funded without federal aid tend to be less complex, constructed to lower geometric design standards, and more often involve resurfacing without widening or other geometric improvements. In a number of states, however, state-funded RRR projects are generally of the same scope as those funded using federal aid. Generally, RRR projects entail costs in the following work categories:

- *Site preparation and earthwork:* clearing, tree removal, excavation, placement of new embankment, and grading;
- *Drainage:* construction of ditches, drains, culverts, and other minor structures required for drainage;
- *Pavement:* all pavement construction—subbase, base, and top course—on lanes and shoulders;
- *Structures:* rehabilitation or replacement of bridges and larger culverts;
- *Traffic and safety:* placement of permanent traffic control devices, lighting, signing, fencing, striping, markers, and similar driver aids;
- *Traffic control:* temporary traffic control measures during construction;
- *Miscellaneous:* other items such as utility pole relocation or curb and sidewalk construction;
- *Right-of-way:* purchase of land or use easements; and
- *Engineering, mobilization, and other:* engineering design and oversight, allowances for contractor front-end costs to assemble necessary equipment and personnel on-site (mobilization), and state sales taxes (if applicable).

For minor widening and resurfacing projects in Washington State, for example, the pavement category accounts for 45 percent of all costs exclusive of right-of-way (Table 4-1). Engineering, mobilization, sales tax, and contingency allowances account for an additional 24 percent, with the remaining costs divided equally among the other categories. In Florida, however, where the terrain is much flatter, the pavement category accounts for an even greater share of project costs (76 percent).

TABLE 4-1 Percent of RRR Costs by Category for Typical Resurfacing and Minor Widening Projects

Category	Washington	Florida	Illinois
Site preparation and earthwork	9	1	10
Drainage	9	2	5
Pavement	45	76	65
Structures	–	–	–
Traffic and safety; traffic control	11	8	5
Miscellaneous	2	–	6
Engineering, mobilization, and other	24	13	9
Total	100	100	100
Shoulder type	Paved	Paved	Unpaved

NOTES: Right-of-way is excluded. Category includes highway with 11-ft lanes and 3-ft shoulders widened to 12-ft lanes and 4-ft shoulders.

Federal-Aid Project Costs

In the 15 case study states, the cost of resurfacing and minor widening projects (including resurfacing without widening) using federal aid ranged from $73,000 to $557,000/mi, almost an eightfold difference (Table 4-2). This

TABLE 4-2 Resurfacing and Minor Widening Construction Costs for Different Funding Sources

State	Federal Aid (cost/mile, $thousands)	State Funds Only (cost/mile, $thousands)	
		Medium and Thick Overlays[a]	Thin Overlays and Seal Coats[b]
Arizona	222	89	19
California	202	198	–
Florida	156	130	–
Illinois	225[c]	206	–
Michigan	196	148	–
Mississippi	122	–	11
Missouri	99	–	11
New Hampshire	477	–	17
New Jersey	430	–	–
Ohio	73	–	47
Texas	557	–	35
Virginia	276	–	–
Washington	–	–	20–48

NOTE: Dashes indicate data not available.
SOURCE: Compiled from state highway agency fiscal year construction programs; fiscal years vary from 1983 to 1985; costs generally exclude right-of-way acquisition and engineering.
[a] Overlay thickness usually greater than 3/4 in.
[b] Overlay thickness usually less than 3/4 in.
[c] Excludes Interstate transfer project in Chicago area.

variation results from many of the factors noted previously—including differences in topography, scope of geometric and roadside improvements, and usual pavement and shoulder design practices.

Urban system RRR projects are generally the most costly on a per-mile basis, followed by primary system projects, with secondary system projects costing less than the other two. This sequence holds in five of the six case study states for which comparable data are available for all three systems (Table 4-3). The exception is New Jersey, where costs for primary system projects exceed those for urban systems, possibly because the primary system is heavily urban (nearly 50 percent) in New Jersey.

TABLE 4-3 Resurfacing and Minor Widening Construction Costs for Different Federal-Aid Systems

State	Average Cost per Mile ($ thousands)			
	Primary	Secondary	Urban	Combined
California	203	173	237	202
Michigan	195	138	229	196
Missouri	95	87	150	99
New Jersey	603	243	437	430
Ohio	83	54	92	73
Texas	521	328	1,255	557

NOTE: Federal-aid projects only.
SOURCE: Compiled from state highway agency fiscal year construction programs; fiscal years vary from 1983 to 1985; costs generally exclude right-of-way acquisition and engineering; for case study states reporting data for all three systems. Includes some nonfreeway, multilane projects.

Non-Federal-Aid Project Costs

Resurfacing and minor widening projects funded without federal aid are less costly on average than those funded with federal aid. Average costs reported for such projects in five of the case study states range from $89,000 to $206,000/mi, or between 40 and 98 percent of the average per-mile costs of projects funded with federal aid (Table 4-2).

Some states also repair pavement surfaces by using thin overlays (usually less than ¾ in. thick) or seal coats, both of which are ineligible for federal aid and therefore must be funded exclusively from state or local sources. Such projects usually involve little work and cost only about one-tenth as much as federal aid resurfacing and minor widening projects.

ADDED PROJECT COSTS FOR GEOMETRIC IMPROVEMENTS

The figures reported in the preceding section illustrate average costs of RRR projects and the variation of costs by state and highway system. These data do not indicate the proportion of RRR spending related to geometric improvements, nor do they indicate how RRR costs might be affected by design standards that require changes to existing highway geometry.

To explore the safety cost-effectiveness of design standards, such information is required, ideally as relationships between incremental project costs and incremental changes in highway geometry that can be generally applied in all RRR project situations. However, for reasons summarized earlier in this chapter no such generally applicable relationships exist. Therefore, studies of cost and safety trade-offs must rely on more limited approaches to cost estimation. Usually cost relationships are tailored to a particular state's experience; the relationships may not be directly applicable to any given project but nevertheless illustrate the added project costs that are incurred when specific geometric improvements are made.

Researchers use two general approaches for developing such relationships. The first, an engineering approach, assumes hypothetical project conditions and required improvements, estimates the construction quantities (e.g., cubic yards of excavation, tons of asphalt) needed to make the improvement, and translates these quantity estimates into costs using typical unit prices from earlier contract experience. The second approach, a more statistical one, relies on records from a sample of actual projects and attempts to relate the variation in costs between projects to differences in project scope, including geometric improvements. The engineering approach clearly links costs to specific design differences, but the hypothetical conditions selected may or may not be representative. The statistical approach, on the other hand, is more broadly representative of actual project experience, but does not have such a clear, reproducible link between specific design features and cost. To estimate relationships between RRR project costs and geometric design, both approaches were used in this study. Where reliable data on project costs and scopes were available for certain features, statistical relationships were used. Elsewhere the study relied on illustrative engineering estimates.

RRR project costs are summarized next for resurfacing-only projects and for the added costs of lane and shoulder widening, flattening sideslopes, removal of roadside obstacles, bridge widening, horizontal curve reconstruction, and lengthening crest vertical curves.

Resurfacing Only

For projects that involve only resurfacing and no geometric improvements, sufficient data are generally available to relate typical project costs within a state to the repaving width. For two-lane rural roads, such relationships are available for the three case study states *(2-4)* in the systemwide cost-effectiveness analyses presented in Chapter 5. Similar relationships have been developed in Kentucky *(5)* and Wisconsin *(6)*; data from Kentucky are presented here along with data from the three case study states (Table 4-4).

Because topography and terrain have little or no effect, resurfacing-only project costs are reasonably consistent for the four states, ranging from a low of $102,000/mi in Illinois to a high of $134,000/mi in Florida for a two-lane highway with 11-ft lanes and no shoulders (Table 4-4). The 30 percent spread between the states is related primarily to differences in pavement unit costs

TABLE 4-4   Resurfacing and Incremental Widening Costs — Two-Lane Rural Highways *(2-6)*

State	Added Cost per Mile ($thousands)
Resurface Two 11-ft Lanes, No Shoulders	
Florida	134
Illinois	102
Kentucky	105
Washington	122
Widen Lanes 1 ft in Each Direction	
Florida	20
Illinois	32
Kentucky	38
Washington	44
Widen Paved Shoulders 1 ft in Each Direction	
Florida	14
Illinois	10
Kentucky	18 (includes some dense graded)
Washington	21
Widen Unpaved Shoulders 1 ft in Each Direction	
Florida (sodding)	4
Illinois (crushed stone)	4
Kentucky (earth or turf)	14
Washington (crushed stone)	8 (new construction)

NOTES: Costs include allowances for engineering (7 percent) and mobilization and minor roadside, intersection, and safety improvements typically a part of a resurfacing project. Costs are adjusted to 1983 levels using FHWA highway construction cost indices. Right-of-way acquisition costs are excluded.

and pavement overlay thicknesses (always greater than ¾ in. and usually in the 1.5- to 3.0-in. range).

Lane and Shoulder Widening

The added costs for lane and shoulder widening on resurfacing projects vary more than the costs for resurfacing only because of the effects of terrain and topography. In the four states given in Table 4-4, the added costs for widening lanes by 1 ft in each direction ranges from $20,000/mi in Florida to $44,000/mi in Washington State, more than a twofold difference. Widening paved shoulders costs about one-half as much as widening lanes because widening at the shoulder edge requires less complex construction methods and thinner pavements are used. If shoulders are widened but left unpaved, shoulder widening costs are reduced by about 50 percent (Table 4-4).

Flattening Sideslopes

Flattening sideslopes at spot locations on RRR projects is relatively common, but the costs vary widely depending on the scope of the sideslope improvement and site conditions. Earthwork is the principal cost component of sideslope flattening and involves excavation (on or off the site), transport, placement, and compaction of earth to build up slopes on fill sections (see Figure 4-1). Flattening sideslopes on cut sections involves excavating the existing sideslopes and removing the excavated soil or rock.

Earthwork unit prices for excavation, transport, placement, and compaction range from $2 to $12/yd^3 depending on the total quantity involved (larger quantities have lower unit prices), source of the material (on or off-site), amount of transport involved, and the nature of the material (rock, clay, loam,

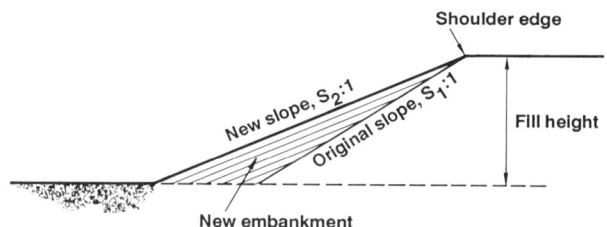

FIGURE 4-1 Flattening sideslopes on fill cross sections.

etc.). To illustrate the sensitivity of sideslope flattening costs to the slope geometry of fill sections, the sideslope flattening costs were compared (i.e., added RRR project costs in terms of cost per mile) assuming earthwork at $7/yd^3. These added costs are moderately sensitive to the original and new slopes, but are very sensitive to the fill height (Table 4-5). For example, flattening a 2:1 original slope to 4:1 costs $9,000/mi with a fill height of 2 ft and $34,000/mi with a fill height of 4 ft.

TABLE 4-5 Illustrative Costs for Flattening Sideslopes on Fill Sections

Original Slope	New Slope	Construction Cost per Mile ($thousands) for One Side of Highway			
		Height = 2 ft	Height = 4 ft	Height = 6 ft	Height = 8 ft
1:1	2:1	5	17	37	66
	3:1	9	33	74	130
	4:1	13	50	110	194
2:1	3:1	5	18	38	67
	4:1	9	34	74	131
	6:1	18	67	147	259
3:1	4:1	5	18	39	68
	6:1	14	51	112	196

NOTES: Based on an earthwork unit price of $7/yd^3; seeding at $1,000/acre; drainage at 15 percent of earthwork costs; plus 25 percent for engineering, mobilization, and miscellaneous expenses. No right-of-way costs are included.

Removal of Roadside Obstacles

Extending the width of the recovery zone beyond the roadway surface can lead to improved highway safety. To accomplish this, objects can be removed, relocated, or protected. The associated costs of these actions have been well-documented in past studies and by several state highway agencies (Table 4-6). Guardrail installation costs range from about $10 to $20/linear ft with no apparent economies of scale. Tree removal, however, does exhibit rather high fixed costs; the unit cost of removing only a few trees is quite high. The costs of relocating signs and utility poles are highly variable, in part because right-of-way requirements differ from site to site.

Bridge Widening

As for many other types of improvements, the added costs for bridge widening can be estimated from statistical studies of past construction costs or from

TABLE 4-6 Unit Costs for Selected Roadside Obstacle Removal and Protection Strategies *(7-10)*

Type of Action	Unit Cost (1985 $)	Source	Remarks
Guardrail			
Removal	1.65/linear ft	*(8)*	
	2.86/linear ft	Washington State Department of Transportation	
Installation	11.00/linear ft	*(8)*	
	17.20/linear ft	Montana Department of Transportation	Additional $733 for each end treatment
	10.00/linear ft	*(7)*	
	21.16/linear ft	Washington State Department of Transportation	
Replacement	25.00/linear ft	New York State Department of Transportation	Includes removal of old rail and some bridge rail
Bridge rail end treatment	6,300/bridge	Montana Department of Transportation	New rail costs same as guardrail
	5,280/bridge	*(7)*	
Tree removal	238/tree	New York State Department of Transportation	Based on removal of 184 trees on a single project
	264/tree	*(7)*	
	220/tree	*(8)*	
	660/tree	*(9)*	Less than 100 nonmarketable trees removed
	85/tree	*(9)*	More than 100 nonmarketable trees removed
Utility pole relocation			
Wood/telephone	405/pole	*(10)*	Rural
Wood/low power	1,490/pole	*(10)*	Rural
Wood/high power	2,660/pole	*(10)*	Rural
Nonwood	2,040/pole	*(10)*	Rural
All types	1,585/pole	*(7)*	
	2,580/pole	*(9)*	
Sign relocation	195/sign	*(8)*	"Small" sizes
Breakaway sign installation	210/sign	*(9)*	
Impact attenuator installation	4,400/unit	*(8)*	Sand-filled type

NOTE: 1985 costs calculated using FHWA composite construction indices.

engineering analyses. Statistical data are commonly used for planning studies, and widening costs are usually expressed in terms of dollars per square foot. The influences of site conditions on the costs of many widening projects are minimal and are not considered in the analysis. Because widening of existing bridges has not been a common feature of RRR projects, cost estimates must rely on data collected for other types of project funding.

Typical costs (in 1985 dollars) for widening existing structures and installing new bridge rail in Washington State ranged from $125/ft^2 for "minimal" widening (up to 5 ft) to $87/ft for "moderate" widening (10 or more ft). This range suggests the existence of a fixed-cost component not influenced by the amount of widening, such as the costs of installing bridge rail, and possibly economies of scale as well. An inverse relationship, capturing these effects and showing reduced sensitivity of unit construction costs to each foot of widening as the total amount increases, was fit to these two points and yielded the following cost model:

$$C = 69 + \frac{225}{W} \tag{1}$$

where C equals the unit cost of bridge widening in dollars per square foot and W equals the width to be added to the bridge (in feet). In calibrating this expression, minimal widening was assumed to be 4 ft and moderate widening, 12 ft. A project that involved widening a bridge 8 ft would thus be expected to cost about $97/ft^2.

Other cost estimates obtained from the literature *(7, 11)* are somewhat lower but reflect the variability expected among the states. These per-square-foot estimates, when factored to 1985 dollars, were $45 for Colorado, $145 for Illinois, $70 for New York, and $85 for Virginia *(11)*. Later, other researchers estimated bridge widening costs to range from about $66 to $82/ft^2, again expressed in 1985 dollars *(7)*. Although on the high side, the Washington State model as expressed by Equation 1 is not substantially different from other estimates and is favored because of the sensitivity it expresses between unit construction costs and the amount of widening.

Many existing highway bridges cannot be widened because of the bridge type (e.g., through trusses) or structural inadequacy. If additional bridge width is required, the only remedy is to remove the existing structure and replace it with a new, wider one. Using data from Washington State (expressed in 1985 dollars), estimated costs for such an alternative are as follows:

	$Cost/ft^2$ ($)
Remove existing structure	6
Construct new structure	
Prestressed concrete girder	56
Concrete posttensioned box girder	69
Steel plate girder	106
Short span concrete slab	56

Horizontal Curve Reconstruction

Flattening a horizontal curve, which increases its length, usually requires full reconstruction between the beginning and ending points of the curve. Thus, while RRR projects generally do not entail reconstruction, it will be required for a portion of the project length if a curve is to be flattened.

Little has been published about the costs of reconstructing horizontal curves. Because curve reconstruction is relatively uncommon and costs are highly dependent on site conditions, state highway agencies generally do not accumulate enough experience to statistically relate average curve reconstruction costs to changes in curve geometry. Hence, data on typical relationships between cost and curve geometry based on completed projects are not available.

Nevertheless, the cost to flatten a horizontal curve is related to the original curve geometry (central angle and degree of curvature) and the change in that geometry [i.e., degree of curvature of the reconstructed curve (see Figure 3-11, Chapter 3)]. In general, reconstruction costs increase as either the curve central angle, the original degree of curvature, or the change in degree of curvature increases. At a particular curve on an existing roadway (with a given central angle and initial degree of curvature), the cost to reconstruct the curve will increase as the change in degree of curvature increases.

To illustrate the order of magnitude of reconstruction costs and the sensitivity to curve geometry, a cost relationship was developed for hypothetical conditions using Washington State unit costs (Appendix I). For a central angle of 50 degrees and initial degree of curvature of 15 degrees, this relationship predicts that it will cost about $150,000 to reduce the degree of curvature to 10 degrees. To reduce it to 6 degrees, the cost increases to about $225,000 (Figure 4-2). In this case, the initial 15-degree curve would accommodate a maximum design speed of 36 mph, whereas the 10-degree curve would be suitable for a design speed of about 45 mph, and the 6-degree curve would accommodate a design speed of 55 mph. These calculations are based on American Association of State Highway and Transportation Officials' (AASHTO) new construction standards for a superelevation rate of 0.08 *(11)*.

HIGHWAY COSTS AND GEOMETRIC DESIGN   123

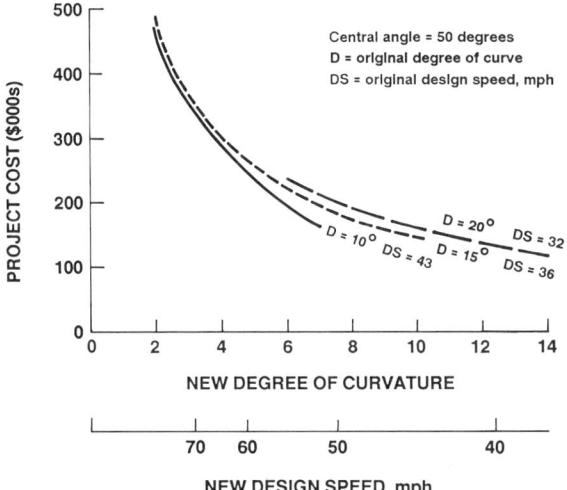

FIGURE 4-2   Illustrative horizontal curve reconstruction costs.

Lengthening Crest Vertical Curves

As is the case with reconstructing horizontal curves, little empirical information is available about costs for improvements to vertical curves. The most common safety-motivated improvement to vertical curves is lengthening crest curves to increase sight distance. Even though such improvements are more common on RRR projects than horizontal curve reconstruction, the practice is still not so common that highway agencies have accumulated enough experience to relate average costs to changes in geometry.

At a given location, however, the cost of lengthening a crest curve will be strongly influenced by existing curve geometry and the changes to that geometry. In general, costs for lengthening a crest curve increase as the change in curve length increases, as the difference between approach and exit grades increases, and as the initial length of the curve decreases (see Figure 3-14, Chapter 3).

To illustrate the order of magnitude of costs involved and the sensitivity to curve geometry, a cost relationship was developed for hypothetical conditions based on Washington State cost data (Appendix I). As an example of cost effects, consider a 200-ft curve at a location where the difference between approach and exit grades is 4 percent. The maximum design speed for such a curve is 35 mph (Table 4-7). According to the cost relationship, lengthening the curve to 400 ft with a maximum design speed of 43 mph would cost about $68,000. Lengthening it to 800 ft, which would allow a maximum design speed of 53 mph, would roughly double the cost to $136,000. If the grade difference was 6 percent instead of 4 percent the costs would be about 15 percent higher.

TABLE 4-7 Illustrative Costs for Lengthening Crest Vertical Curves to Increase Stopping Sight Distance

	Initial Curve		Improved Curve		
Difference in Grades %	Length, ft	AASHTO Design Speed mph[a]	Length, ft	AASHTO Design Speed mph[a]	Project Cost ($thousands)
4	200	35	400	43	68
4	200	35	800	53	136
4	400	43	800	53	126
4	400	43	1,000	57	165
4	600	48	800	53	109
6	200	31	800	47	157
6	200	31	1,400	56	281
6	400	38	800	47	144
8	600	39	1,000	46	198
8	600	39	1,800	55	399

NOTES: Costs are based on unit prices from Washington State for two-lane rural roads (see Appendix I); schematic of vertical curve geometry is shown in Figure 3-14, Chapter 3.
[a] AASHTO maximum design speed based on stopping sight distance criteria (12, p. 308).

Intersection Improvements

Costs for intersection improvements are highly variable, depending on the physical and operational features to be improved and other site-specific conditions. Representative costs for widening and new channelization of an existing intersection are $100,000 to $150,000 (Table 4-8). Improvements such as the construction of new turning lanes and realignment of curbs are less costly—typically $10,000 to $20,000 per intersection. Also, some intersection improvements, such as rechannelization using pavement markings and upgraded traffic control devices, can be implemented for much less than $10,000 per intersection as part of a RRR project.

TABLE 4-8 Representative Costs of Intersection Improvements *(14)*

Type of Project	Construction Cost (1983 $)
Widening and new channelization	100,000-150,000
Installation of new traffic signals	60,000-100,000
Reconstruction of one approach	50,000-75,000
Construction of new turning lanes	10,000-20,000
Realignment of curb	10,000

RIGHT-OF-WAY REQUIREMENTS

As noted in Chapter 2, most federal-aid RRR projects do not involve right-of-way acquisition. Existing rights-of-way are often large enough to accommodate minor lane and shoulder widening and some roadside improvements. Of the 15 state highway agencies visited for this study, only Virginia routinely acquired additional right-of-way for RRR work. In Virginia, right-of-way acquisition is necessary on some secondary highways because narrow existing rights-of-way are common, sometimes as narrow as 30 ft.

Although many geometric improvements can be completed within existing right-of-way, others cannot. Lane and shoulder improvements occasionally require additional right-of-way as do roadside improvements such as slope flattening or removal of trees and other fixed objects. Reconstruction of horizontal curves almost always requires additional right-of-way.

State highway agencies generally resist acquiring right-of-way for RRR projects because right-of-way acquisition will be costly and time consuming, or it will adversely affect aesthetics and the community.

Financial Costs

The financial costs of right-of-way acquisition include not only land purchase prices, but also administrative costs associated with negotiation and condemnation. As a result, right-of-way costs can include a fixed per-parcel component and a variable component based on parcel size and location. Washington State, for example, reported a fixed cost of approximately $10,000/parcel and a variable component of approximately $5,000/acre in rural areas and $50,000 to 100,000/acre in urban areas.

Time Costs

Right-of-way acquisition often requires considerable time, in some cases 2 years or more. In Florida the state highway agency allows 18 months for land acquisition. Such a requirement conflicts with RRR project schedules, which are usually geared toward providing urgent pavement repair within one year. In many states the delay in project scheduling is as important as financial costs for avoiding right-of-way purchases for RRR improvements.

Aesthetic and Community Impacts

Removing trees or taking portions of homeowners' front yards is unpopular and can arouse community opposition to RRR improvements. Although these possibilities arise more often in urban and suburban settings, they occur in rural areas as well. Some state highway agencies have issued specific policies on circumstances warranting tree removal. Geometric improvements on RRR projects are rarely foregone because of negative aesthetic and community impacts alone; however, in conjunction with financial and time costs they form an additional barrier to right-of-way purchases.

MAINTENANCE COST IMPLICATIONS

The effect of geometric improvements on maintenance costs and requirements has been considered in only a general way in a few prior studies. State maintenance records generally have been underutilized, and it is only since the advent of pavement management systems that they have been used to relate maintenance activities and costs to pavement condition. Future use of these records may lead to a better understanding of the effects of geometric design on maintenance costs, but in the meantime no reliable quantitative relationships are available that can estimate added maintenance costs that are a result of geometric improvements. Nonetheless, improvements that directly affect the amount of roadway surface can be expected to increase maintenance costs. Other improvements, such as sideslope flattening and widening roadside clear zones, will not have a substantial effect on maintenance requirements.

Maintenance requirements can be expected to increase as highways are widened and as more surface area must be maintained. Although some economies of scale can be realized in maintaining the additional area, they are probably small, so that as an approximation, lane and shoulder maintenance costs are roughly proportional to surface area.

The added maintenance costs that result from cross-section improvements are small relative to either the total maintenance costs or the capital costs of widening. For example, consider the routine maintenance costs on the following hypothetical RRR project:

Existing Conditions

Length = 2.0 mi
Lane width = 11 ft
Shoulder width = 3 ft
Rolling terrain
Paved shoulders

Required Improvements

Lane width = 12 ft
Shoulder width = 4 ft

Construction costs for hypothetical RRR project are as follows:

Improvement	Cost and Cost/ft-mi ($)
Lane widening	89,200 (22,300)
Shoulder widening	42,400 (10,600)
Resurfacing	280,000 (5,000)
Total	411,600
Annual construction costs (7 percent discount rate over a 30-year project life)	33,200

Annual routine pavement maintenance costs at $250/ft-mi are as follows:

No Improvement ($)	With Widening ($)	Increase ($)	% Increase	Increase as Percent of Annualized Construction Cost
14,000	16,000	2,000	14.3	6.0

NOTE: Construction unit costs were derived empirically from case study state project data. Pavement maintenance costs are from the Washington State Pavement Management System.

Although maintenance costs on this segment will increase 14 percent because of additional surface area, this increase represents less than 10 percent of annualized project construction costs.

Data from the Highway Performance Monitoring System (HPMS) were used to obtain a rough estimate of the nationwide effect of cross-section improvements on maintenance costs. These estimates are given for two-lane rural federal-aid highways:

128 DESIGNING SAFER ROADS

	Existing Surface (ft-mi)[a]	Added Surface Area If Improved (ft-mi)[b]	Percent Increase
Lanes	11,681,000	432,000	3.7
Shoulders	5,146,000	150,000	2.9
Total	16,827,000	582,000	3.5

	1984 Maintenance Cost ($thousands)[c]	Cost Increase Resulting From Improvement ($thousands)
State	1,329,773	46,542
Local	629,063	22,017
Total	1,958,836	68,559

Source: Analysis of 1983 Highway Performance Monitoring System data base and *Highway Statistics 1984*.

[a] Average lane width = 10.9 ft; average shoulder width = 4.8 ft; total mileage = 536,415.
[b] If 1978 FHWA RRR standards are applied.
[c] Includes costs for facilities, structures, and snow and ice control only.

State and local governments spend nearly $2 billion on maintenance activities on two-lane rural federal-aid highways *(13)*. If lanes and shoulders of these highways were to be widened to the special RRR standards proposed by FHWA in 1978, an increase of 3.5 percent would have to be maintained in the surface area. Assuming that maintenance costs rise proportionally with surface area, an extra $68 million would have to be expended nationwide. However, because cross-section improvements would be made over a period of time, maintenance costs would increase only slightly each year; maintenance costs would increase by approximately $3.4 million/year over 20 years.

Maintenance costs can be an important consideration in choosing among alternative roadside improvements. For example, flattening sideslopes becomes more attractive as an alternative to placing guardrail when the costs of maintaining the guardrail are taken into account.

Alignment improvements can also affect highway maintenance costs. For example, alignment improvements that require pavement reconstruction at curves will reduce maintenance costs in the short term because newer pavements cost less to maintain.

SUMMARY

Geometric design improvements on RRR projects affect both the initial capital investment required for the project and future maintenance requirements. Although initial project costs tend to be the dominant consideration for highway agencies making decisions about geometric design improvements, maintenance costs can also be important, and neglecting to take them into account can result in poor design decisions.

The added cost for specific design improvements was found to vary widely from project to project because of variations in site conditions, labor and materials costs, design practices, and project scale.

In the case study states, federal-aid resurfacing and minor widening project costs ranged from $73,000 to $557,000/mi. RRR projects on urban systems are generally the most costly, followed by primary and secondary system RRR projects. Resurfacing and minor widening projects funded without federal aid are, on average, less costly than those funded with federal aid, although in some states the difference is small.

Unit costs were assembled for resurfacing-only projects and for lane and shoulder widening. Differences in resurfacing costs among states were the result of differences in unit costs for materials and the thickness of typical overlays. Larger state-to-state variations were found in the unit cost for lane and shoulder widening because of the effects of terrain and topography.

Unit costs were assembled for flattening sideslopes and removing roadside obstacles. Sideslope costs were found to be highly sensitive to fill heights. Costs for both types of improvements vary considerably depending on whether acquisition of right-of-way is required.

Rough, quantitative relationships were developed for estimating costs for reconstructing horizontal and vertical curves and widening bridges. The relationships account for economies of scale in these improvements and for the considerable effects of central angle (horizontal curves) and grade (crest curves) on costs.

REFERENCES

1. *Dodge Guide to Public Works and Heavy Construction Costs.* McGraw-Hill, New York, 1985.
2. Unpublished cost estimates from Long Range Cost Estimation Procedure, Florida Department of Transportation, Tallahassee, 1985.
3. Unpublished data from a sample of RRR projects provided by Washington State Department of Transportation, Olympia, 1985.
4. Unpublished cost estimates provided by Illinois Department of Transportation, Springfield, 1986.

5. C. V. Zegeer, R. C. Deen, and J. G. Mayes. "Effect of Lane and Shoulder Widths on Accident Reduction on Rural Two-Lane Roads." In *Transportation Research Record 806*, TRB, National Research Council, Washington, D.C., 1981, pp. 33–43.
6. *An Evaluation of Alternate Shoulder Width Standards for New Construction on Wisconsin's Two-Lane State Truck Highways.* Summary Report. Wisconsin Department of Transportation, Madison, Feb. 1984.
7. S. A. Smith, et al. *Identification, Quantification and Structuring of Two-Lane Rural Highway Safety Problems and Solutions.* Report FHWA/RD-83/022. FHWA, U.S. Department of Transportation, June 1983.
8. *Cost-Effective Cross-Section Design for Two-Lane Roads.* FHWA, U.S. Department of Transportation, (forthcoming).
9. J. L. Graham, and D. W. Harwood. *NCHRP Report 247: Effectiveness of Clear Recovery Zones.* TRB, National Research Council, Washington, D.C., May 1982.
10. C. V. Zegeer and M. R. Park, Jr. *Cost-Effectiveness of Countermeasures for Utility Pole Accidents.* FHWA, U.S. Department of Transportation, Jan. 1983.
11. J. C. Glennon. *NCHRP Report 148: Roadside Safety Improvement Programs on Freeways: A Cost-Effectiveness Priority Approach.* TRB, National Research Council, Washington, D.C., 1974.
12. *A Policy on Geometric Design of Highways and Streets.* American Association of State Highway and Transportation Officials, Washington, D.C., 1984.
13. *Highway Statistics 1984.* FHWA, U.S. Department of Transportation.
14. T. R. Neuman. *NCHRP Report 279: Intersection Channelization Design Guide. 3.* TRB, National Research Council, Washington, D.C., 1985.

5
Safety Cost-Effectiveness of Geometric Design Standards

The safety cost-effectiveness of design standards was analyzed for resurfacing, restoration, and rehabilitation (RRR) projects on the basis of relationships between safety and geometric design (Chapter 3) and the relationships between cost and geometric design (Chapter 4). The principal emphasis in this chapter is on the safety effects of design improvements. However, operational effects, such as user travel time and operating cost savings as a result of design improvements, are also considered.

Cost-effectiveness analyses were conducted at both the project and system level. The objective of the project-level analyses was to establish the circumstances under which a given design improvement is cost-effective by examining cost versus safety trade-offs for a series of representative projects. This information can be used in setting minimum standards or establishing design practices so that opportunities for cost-effective improvements are carefully considered by designers.

Systemwide (or statewide) safety gains and costs of geometric improvements were also considered in order to address the overall financial and budgetary consequences of design standards and the trade-off between pavement condition and safety that is central to the debate over RRR standards. In essence, the systemwide approach addresses two questions: From a safety standpoint, are certain kinds of geometric improvements to existing highways worth making throughout a highway system? At current funding levels, can state highway agencies afford these improvements while maintaining pavements in reasonable condition?

EARLIER STUDIES OF SAFETY COST-EFFECTIVENESS IN HIGHWAY DESIGN

Existing highway design standards—both for new construction and for RRR work—generally are not linked to explicit assessments of cost-effectiveness. Although the standards reflect judgments about the effects of changes in design variables on safety, traffic operations, construction cost, and other considerations, they are not the products of formal cost-effectiveness studies in which these effects are quantified and trade-offs are analyzed. State highway agencies, the American Association of State Highway and Transportation Officials (AASHTO), the Federal Highway Administration (FHWA), and others who adopt standards or recommend design guidelines have not relied on cost-effectiveness analyses for several reasons:

• Surprisingly little is known about how much accident rates will decline with improvements in road design. Researchers have often reached conflicting conclusions, and explicit, widely accepted quantitative relationships are not available from the literature (Chapter 3).
• The cost effects of design standards vary widely because of differences in site conditions, unit costs, design practices, and project scale (Chapter 4).
• Conclusions from a cost-effectiveness analysis can be very sensitive to discount rates, imputed accident costs, imputed value of time, and other assumptions made to facilitate comparisons of benefits and costs.
• Even if the cost-effectiveness analyses produce the results intended, questions remain about how these results should be used in setting standards. What other factors should be considered in addition to cost-effectiveness, and what are the relative weights of these factors? How should the underlying uncertainties of cost-effectiveness analysis be taken into account?

Because of the preceding problems, researchers who have analyzed the safety cost-effectiveness of highway improvements generally have sought to provide designers with methods that can be used to produce more cost-effective designs, rather than a better basis for setting standards *(1-7)*.

Leisch and Neuman *(8)* explicitly examined the cost-effectiveness of new construction standards for locally administered highways in Minnesota. They argued that design standards are "... one of the most important tools the highway engineer has ..." and "... limited construction and maintenance budgets and increased public pressure on the engineer to justify the expenditure of public highway funds demand these standards to be cost-effective."

Other researchers have presented the development of cost-effective designs for individual projects as an alternative to designing to standards. For example, Jorgensen Associates *(1)* describe the objective of their study as developing a methodology for "tailoring the designs of individual projects rather than developing designs through rigid application of design standards." Similarly, Graham and Harwood *(3)* state "the cost-effectiveness of roadside design improvements can vary widely between highway sections based on accident rates, traffic volumes, terrain, required construction quantities, unit construction costs, and right-of-way requirements. There is a clear need for a roadside design process based on cost-effectiveness considerations rather than a single, fixed roadside design policy."

The question of customized design (versus reliance on standards) depends on the extent to which safety, cost, and other impacts of a given design improvement vary from site to site. Customized designs are appropriate when site-to-site variations are large and the factors influencing these variations are not easily represented in design standards. Standards are appropriate when little site-to-site variation exists or when factors influencing these variations (e.g., traffic levels) are easily taken into account when applying the standard. In any case, it is likely that design standards will remain in some form and it is reasonable to question the cost-effectiveness implications for highway systems as a whole.

To the extent that the authors of the preceding studies examined or commented on standards, they did so with respect to new construction standards rather than RRR standards. Both Jorgensen Associates *(1)* and Leisch and Neuman *(8)* noted the strong sensitivity of cost-effectiveness to traffic volume and implied that new construction design standards might be too stringent for low-volume rural highways and perhaps not stringent enough for nonfreeway highways with high traffic volumes.

SCOPE AND FRAMEWORK OF COST-EFFECTIVENESS ANALYSES

The principal measure of cost-effectivness used in this study is added RRR project cost per accident eliminated. Improvements to geometric design for a given RRR project generally reduce the number of accidents and increase the cost of the project. Cost per accident eliminated shows the added cost required to eliminate one accident.

Cost per accident eliminated is calculated as follows:

- Estimate the change in accident rate for the design improvement under consideration and, given average daily traffic (ADT), the change in the number of accidents per year.
- Estimate the added cost required to implement the design improvement as part of a RRR project.

- Annualize the added cost, based on an assumed project life and discount rate.
- Calculate cost per accident eliminated as added cost divided by accidents eliminated.

For simplicity, cost-effectiveness analyses presented in this chapter assume constant traffic volumes expressed in terms of average daily traffic.

As calculated in the preceding steps, cost per accident eliminated does not include the benefits to highway users associated with travel time and operating cost savings. These benefits can be accounted for in the cost per accident eliminated framework by subtracting them from the cost required to implement a design improvement and calculating net (construction minus user) cost per accident eliminated. However, a shortcoming of this method is that if user savings exceed the cost to implement the improvement, the net cost per accident eliminated is negative and therefore not meaningful.

The advantage of using cost per accident eliminated as the measure of safety cost-effectiveness is that it focuses directly on the safety versus cost trade-off and allows designers or policymakers to impute their own values to accidents eliminated. To use this information, however, users must understand the underlying distribution of accidents by severity (e.g., for the type of accident under consideration, how many fatalities and injuries are there per thousand accidents?). A disadvantage of this approach is that it understates safety benefits when design improvements act principally to reduce the severity of accidents, rather than the number of accidents, and overstates safety benefits when design improvements result in higher speeds and increased accident severity. In both cases, cost per fatal or injury accident eliminated may be a more appropriate measure of safety cost-effectiveness.

Past studies of the cost-effectiveness of highway safety improvements differ considerably in the value imputed to eliminating different types of accidents, particularly fatalities (Table 5-1). The National Highway Traffic Safety Administration (NHTSA) uses the future economic production of individuals (the amount of compensation individuals would have received had the fatality not occurred) in assigning cost to a fatality, whereas the National Safety Council (NSC) uses production minus consumption. The "willingness-to-pay" approach uses estimates of the amounts individuals would pay for small reductions in the probability of death.

When the values for different severity classes are weighted together using a typical distribution of accidents by severity for two-lane rural highways, the resulting average value per accident eliminated ranges from $10,000 to $50,000 in 1985 dollars. A specific value is neither recommended nor applied in this study but it is concluded that improvements with a cost per accident eliminated less than $10,000 are clearly justified on safety cost-effectiveness

TABLE 5-1 Alternative Estimates of Accident Costs by Severity

	Cost per Accident ($thousands 1985)[a]		
Severity of Accidents	NHTSA Approach[b]	NSC Approach[c]	"Willingness-to-Pay" Approach[d]
Fatal	394.6	256.5	1,348.7
Injury	11.1	13.2	10.1
Property damage only	1.4	1.2	1.9
All accidents on two-lane rural highways[e]	16.9	13.3	45.3

[a]Costs are updated to 1985 using the implicit price deflator for gross national product.
[b]Costs are based on the 1983 NHTSA report (9). Costs are provided by incident—fatality, injury, or property damage involvement. The unit costs per incident were converted to unit costs per accident using national level summary data on accidents and incidents provided in the report.
[c]NSC costs are based on *Estimating the Cost of Accidents 1984 (10).*
[d]The willingness-to-pay estimates are from Kragh et al. (11).
[e]Costs were calculated using a distribution of accidents by severity from Smith et al. (4). The distribution—3 percent fatal, 37 percent injury, and 60 percent property damage only — is for two-lane rural highways with 400 to 2,000 ADT.

grounds whereas improvements with a cost per accident eliminated greater than $50,000 are not justified solely on safety cost-effectiveness grounds. Improvements with cost per accident eliminated in the $10,000 to $50,000 range may or may not be warranted depending on *(a)* the specific value imputed to accidents eliminated; *(b)* uncertainties surrounding estimates of the added cost for an improvement and the number of accidents it will eliminate; and *(c)* other factors (e.g., environmental effects) that are not accounted for in the calculation of cost per accident eliminated.

The relationship between cost-effectiveness analyses using cost per accident eliminated and the benefit-cost ratio approach employed in other studies is illustrated in Appendix J. In a benefit-cost ratio approach, the analyst imputes dollar values to accidents eliminated. The numerator of the benefit-cost ratio is the sum of annual safety and operational benefits to highway users for the design improvement under consideration. The denominator is annualized construction cost for the improvement.

The benefit-cost approach permits a direct comparison of safety benefits with costs and other consequences of design improvements that can be valued in dollar terms. Further, benefit-cost can account for changes in accident severity because separate unit costs are used for accidents by severity class. Benefit-cost results, however, will be of limited value to users who disagree substantially with the dollar values imputed for accident costs.

Cost per accident eliminated (rather than the benefit-cost ratio) was selected as the principal measure of cost-effectiveness to avoid the arbitrary imputation of dollar values to accidents eliminated. Had a benefit-cost approach been used instead, findings about cost-effectiveness would not change, provided that the dollar value imputed to accidents eliminated falls within the $10,000 to $50,000 range.

Economic Assumptions

For the calculations of cost per accident eliminated presented in this chapter, a discount rate of 7 percent and a project life of 30 years are assumed. The 7 percent discount rate splits the difference between the 4 percent rate recommended by AASHTO *(12)* for low-risk investments and the 10 percent rate recommended by the Office of Management and Budget in Circular A-94. The fixed project life of 30 years was assumed for simplicity. The useful lives of safety improvements vary considerably, and in many cases will be less than 30 years.

For a hypothetical safety improvement with a cost per accident eliminated of $10,000 at a discount rate of 7 percent and a project life of 30 years

- Increasing the discount rate to 10 percent would increase cost per accident eliminated to $13,200 (about 30 percent);
- Decreasing the discount rate to 4 percent would reduce cost per accident eliminated to $7,200 (about 30 percent);
- Increasing project life to 40 years would reduce cost per accident eliminated to $9,300 (about 7 percent); and
- Decreasing project life to 20 years would increase cost per accident eliminated to $11,700 (about 15 percent).

SAFETY-COST TRADE-OFFS

This section contains an examination of safety-cost trade-offs for lane and shoulder widths, horizontal curvature, roadside obstacles, sight distance on vertical curves, and bridge width—the key design features for which quantitative safety relationships have been developed for two-lane rural highways.

Formal analyses of safety-cost trade-offs were not conducted for traffic controls and other low-cost safety measures because of the lack of reliable quantitative relationships for estimating the safety effects of these measures. However, the committee found sufficient information to conclude that such measures can provide significant reductions in the frequency and severity of accidents (Chapter 3). The safety benefits of the measures, coupled with their low costs, are such that the measures can be highly cost-effective on RRR projects.

Lane and Shoulder Widths

Of all the highway geometric features considered in this study, lane and shoulder widths are the most amenable to quantitative analysis. Although the caveats presented about uncertainties in accident relationships and costs apply to lane and shoulder widths, more is known about the safety effects and costs of these features. Also, costs for lane and shoulder widening vary less from site to site than do costs for other features so that findings about their cost-effectiveness are more easily generalized.

Better system-level data are available for existing lane and shoulder designs. Thus, the number of highway miles above and below a given lane and shoulder width standard could be determined because data on these design features are reported in state highway inventories. Such systemwide comparisons between standards and actual conditions are generally not possible for horizontal curves and sight distance on vertical curves because standards for these features are usually expressed in terms of design speeds, and information on design speeds for individual curves is not available in most state highway inventory data sets.

Project-Level Cost-Effectiveness

Important determinants of the cost-effectiveness of lane and shoulder widening are

- Traffic volumes,
- Roadside environment,
- Terrain and highway alignment, and
- Lane and shoulder widths before improvement.

Traffic volumes are a prime consideration in the cost-effectiveness of lane and shoulder widening (Figure 5-1) because the number of accidents eliminated by lane and shoulder widening increases almost in proportion to ADT, whereas costs are not significantly affected by ADT.

Current RRR standards frequently place little weight on ADT. In some states, lane width standards for RRR projects are based solely on design speed. This may result, for example, in widening 10-ft lanes to 11 ft on a highway with 300 ADT but not widening 11-ft lanes to 12 ft on a highway with 5,000 ADT, even though lane widening on highways with high ADT has a much lower cost per accident eliminated.

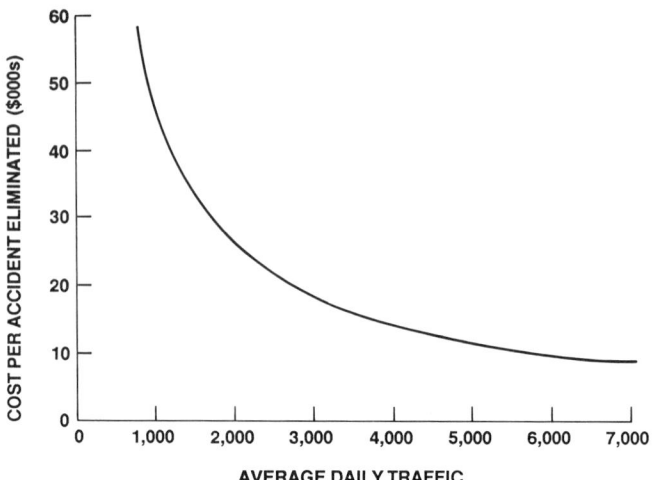

NOTES: Example assumes rolling terrain, 10-ft lanes with 2-ft shoulders before improvement and 11-ft lanes with 4-ft shoulders after improvement. Costs are in 1985 dollars and were calculated using a discount rate of 7 percent and a project life of 30 years.

FIGURE 5-1 Cost-effectiveness of lane and shoulder widening by ADT.

The roadside environment is an important consideration in the cost-effectiveness of lane and shoulder improvements because roadside hazards such as steep slopes and fixed objects affect both the likelihood that an accident will occur and the severity of the accident. In the relationship between accident rates and lane and shoulder width described in Appendix C, the degree of hazard associated with the roadside is represented by a roadside rating ranging from 1 (least hazardous) to 7 (most hazardous). A roadside hazard rating of 1 might involve a flat roadside with no fixed objects within 30 ft of the outside edge of the shoulder. A roadside rating of 7 might involve a steep downward slope or a sheer rock wall along the edge of the shoulder. For highway segments with roadside ratings of 6 or 7, the cost per accident eliminated is less than one-half that for sections with roadside ratings of 1 or 2 (Figure 5-2).

Terrain and highway alignment are also important considerations in the cost-effectiveness of lane and shoulder improvements. Costs for lane and shoulder widening on more rugged terrain generally increase, because of the greater amount of earthwork required. The accident reduction associated with lane and shoulder widening is also affected by terrain and alignment because more rugged terrain and poor alignment means higher accident rates and thus

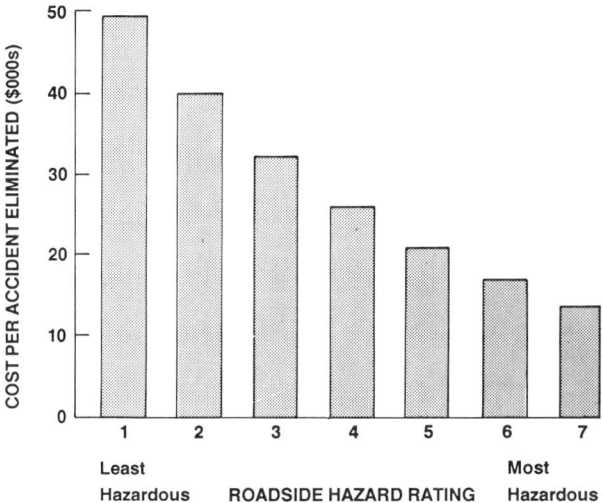

NOTES: Example assumes 2,000 ADT, rolling terrain, 10-ft lanes with 2-ft shoulders before improvement and 11-ft lanes with 4-ft shoulders after improvement. Costs are in 1985 dollars and were calculated using a discount rate of 7 percent and a project life of 30 years.

FIGURE 5-2 Cost-effectiveness of lane and shoulder widening by roadside hazard rating.

the reduction in accidents as a result of a given lane and shoulder width improvement is greater.[1]

The study examined the cost-effectiveness of adding lane or shoulder width using typical unit costs and accident rates for flat, rolling, and mountainous terrain (Table 5-2). On flat terrain, shoulder widening is more cost-effective than lane widening because of the relatively low cost of widening shoulders on flat terrain. On mountainous terrain, however, the cost-effectiveness of lane and shoulder widening are nearly equal.

The cost-effectiveness of lane and shoulder widening diminishes as widths approach the levels mandated by new construction standards (Figure 5-3). Although the cost for widening lanes from 10 to 11 ft is about the same as the cost for widening lanes from 11 to 12 ft, the number of accidents eliminated is less in the latter case.

Lane and shoulder width improvements result in time and cost savings to

---

[1] According to the accident relationship for lane and shoulder width, the percentage improvement in accident rate is constant for given lane and shoulder width improvements. Adding 1 ft of lane width, for example, always causes a 12.1 percent decrease in accidents. For highway sections with poor alignment, the base accident rate is higher so that a given percentage reduction in accidents results in a higher absolute reduction.

TABLE 5-2 Illustrative Cost-Effectiveness of Lane and Shoulder Widening, 2,000 ADT

Terrain	Lanes	Paved Shoulders	Unpaved Shoulders
Added Cost per Mile of Widening (1 ft each direction)			
Flat	37.6	14.0	4.6
Rolling	44.6	21.2	11.8
Mountainous	68.2	44.6	35.2
Accidents Eliminated per Mile per Year			
Flat	0.086	0.058	0.048
Rolling	0.109	0.073	0.061
Mountainous	0.172	0.115	0.097
Cost per Accident Eliminated for Widening			
Flat	35.3	19.6	7.7
Rolling	33.0	23.4	15.5
Mountainous	31.9	31.2	29.3

NOTES: Costs are in thousands of 1985 dollars. Unit costs for widening are based on unit costs for individual states presented in Chapter 4, Table 4-4; to calculate cost per accident eliminated, widening costs were annualized using a 7 percent discount rate and a project life of 30 years; the example assumes 10-ft lanes and 2-ft shoulders before improvement.

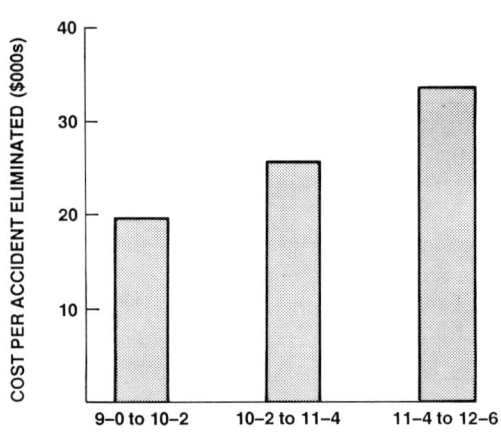

NOTES: Example assumes 2,000 ADT and rolling terrain. Costs are in thousands of 1985 dollars and were calculated using a discount rate of 7 percent and a project life of 30 years.

FIGURE 5-3 Cost-effectiveness of successive lane and shoulder width improvements.

highway users, as well as improved highway safety. Narrow lanes and shoulders on two-lane rural roads cause motorists to drive closer to vehicles in the opposing lane. Motorists must compensate for driving closer to opposing traffic by slowing down and allowing larger headways between vehicles in the same lane. Thus, motorists driving on roads with narrow lanes and shoulders will experience more delay and will drive at lower speeds than on roads with wider lanes and shoulders.

Speed increases as a result of lane and shoulder width improvements offset part of the safety benefit of these improvements because, other things being equal, accident rates increase with speed. This speed-related effect is already accounted for in the accident relationship for lane and shoulder width because the relationship provides estimates of the net effect on safety of width improvements and related speed increases.

The *Highway Capacity Manual (13)* provides factors for adjusting the capacity of two-lane rural highways to account for lane and shoulder width. This study estimated the effects on travel times of improvements to lane and shoulder widths based on the *Highway Capacity Manual* methodology (Appendix K).

At ADT levels less than 2,000, the time savings associated with lane and shoulder width improvements are minimal (Figure 5-4). These savings increase sharply with increasing ADT, however, because the effects of narrow lanes and shoulders are exacerbated when traffic volumes are greater. The

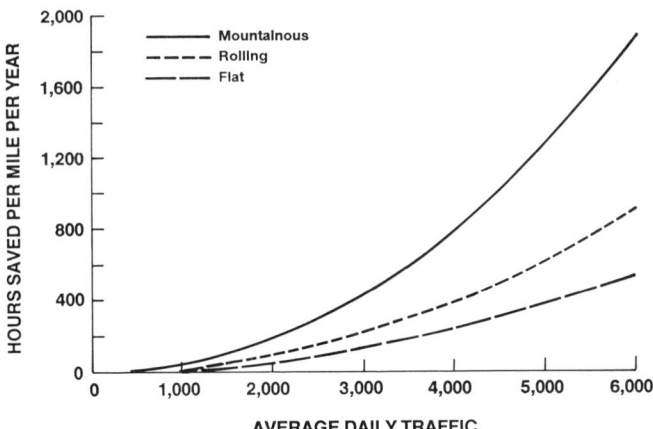

NOTE: Example assumes 10-ft lanes with 2-ft shoulders before improvement and 11-ft lanes with 4-ft shoulders after improvement.

FIGURE 5-4 Travel time savings for lane and shoulder widening.

effect of a given lane and shoulder width improvement on travel time will also vary depending on terrain because curves and grades add to the adverse effects of narrow lanes and shoulders on vehicle operating speeds.

Travel time savings for the first few feet of widening are usually substantially greater than those for further widening (Figure 5-5). For highways with 12-ft lanes and 6-ft shoulders, the time savings for further widening are minimal.

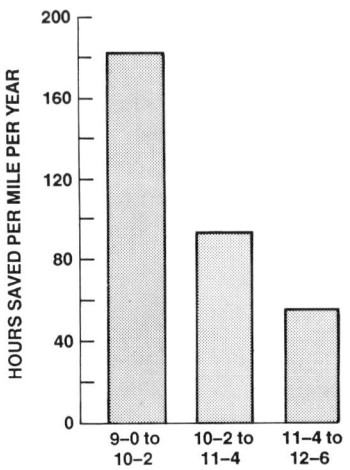

NOTE: Example assumes 2,000 ADT and rolling terrain.

FIGURE 5-5 Travel time savings for successive lane and shoulder width improvements.

Part of the travel time benefit to users associated with wider lanes and shoulders will be offset by the higher operating costs associated with higher travel speeds. This point is illustrated in the following table, which gives the travel time and operating cost effects of an increase in speed from 50 to 55 mph.

User Costs per 1,000 Vehicle Miles

	Travel Time (at $7.50/hr)	Operating Cost (in 1985 $)	Total User Cost ($)
50 mph	150.00	190.72	340.72
55 mph	136.36	195.15	331.51
Difference	13.64	−4.43	9.21

In this example, about one-third of the travel time savings are offset by greater vehicle operating costs.

At greater ADT levels, the net user savings—time savings less added vehicle operating costs—can be substantial in relation to the cost of lane and shoulder widening (Figure 5-6). At 5,000 ADT, for example, the user savings are about 40 percent of the widening cost. At lower ADT levels, however, the user savings are small; for example, less than 2 percent at 1,000 ADT.

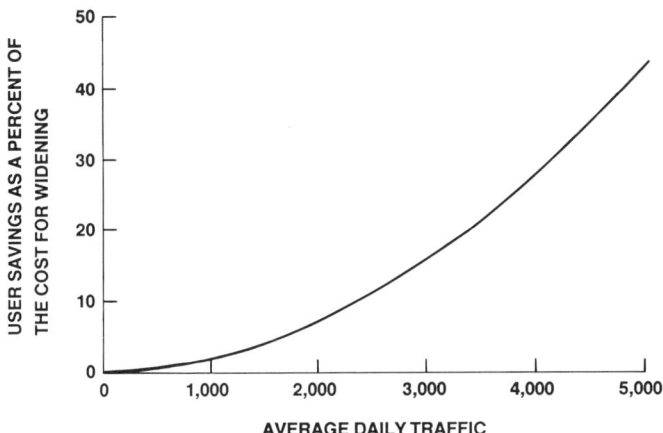

NOTES: Example assumes 10-ft lanes with 2-ft shoulders before improvement, 11-ft lanes with 4-ft shoulders after improvement, and rolling terrain. Travel time savings are valued at $7.50 per vehicle hour.

FIGURE 5-6 User savings as a percent of the cost for lane and shoulder widening.

System-Level Cost-Effectiveness

At the system level, the study examined the cost-effectiveness of alternative minimum lane and shoulder width standards using data on two-lane, rural federal-aid highways from the FHWA's Highway Performance Monitoring System (HPMS), as well as more detailed data from the roadway inventories of Florida, Illinois, and Washington State. The standards are minimum acceptable lane and shoulder widths for a highway segment, given the segment's traffic volume, design speed, percentage of trucks, and highway functional classification. Designers compare the standards with existing lane and shoulder widths to determine whether the existing widths can remain in place or whether they must be upgraded.

Alternative standards examined at the national level include

- Application of AASHTO new construction standards to RRR projects;
- RRR standards developed by AASHTO in 1977 *[Geometric Design Guide for Resurfacing, Restoration, and Rehabilitation of Highways and Streets (14)*, referred to as the AASHTO Purple Book];
- Standards proposed by FHWA in 1978 (but not adopted); and
- A modified version of the 1978 FHWA standards in which the ADT ranges for triggering lane and shoulder width improvements are shifted and wider shoulders are required at the highest ADT levels (Table 5-3). The three alternative standards examined at the national level are described in Appendix L.

TABLE 5-3 FHWA Proposed Standards for Lane and Shoulder Widths With Modifications

Design Year Volume (ADT)	Running Speed[a] (mph)	10 Percent or More Trucks[b]		Less Than 10 Percent Trucks	
		Lane Width	Combined Lane and Shoulder Width[c]	Lane Width	Combined Lane and Shoulder Width[c]
1-750	Under 50	10	12	9	11
	50 and over	10	12	10	12
751-2,000	Under 50	11	13	10	12
	50 and over	12	15	11	14
Over 2,000	All	12	18	11	17

[a]Highway segments should be classified as "under 50" only if most vehicles have an average speed of less than 50 mph over the length of the segment.
[b]For this comparison, trucks are defined as heavy vehicles with six or more tires.
[c]One foot less for highways on mountainous terrain.

The lane and shoulder width standards currently used by states generally fall between the special RRR standards proposed by FHWA in 1978 and AASHTO new construction standards (Chapter 2).

Once it has been determined that a given highway segment must be upgraded because it does not meet minimum standards, state highway agencies will frequently improve the segment beyond the minimum standards up to the levels suggested by AASHTO for new construction (Chapter 2). To estimate accidents eliminated and costs for the four sets of minimum standards less stringent than AASHTO new construction standards, assumptions about the extent of this practice are needed. Because highway agencies are more likely to make improvements beyond the minimum standards when traffic levels are greater, this study assumed for simplicity that segments not meeting minimum standards would be upgraded to *(a)* minimum standards if ADT is 2,000 or less and *(b)* new construction standards if ADT is greater than 2,000. In practice, site-specific considerations such as congestion levels and the roadside environment will govern whether and by how much improved cross sections will exceed minimum standards.

At the national level, the cost for lane and shoulder widening under AASHTO RRR standards is about 20 percent of the cost under AASHTO new construction standards (Table 5-4). The number of accidents eliminated is less than 20 percent, however, so that AASHTO RRR standards perform poorly in terms of cost per accident eliminated. This occurs because, unlike other standards, AASHTO RRR standards do not vary with traffic volume. As shown in the project-level analysis, cost per accident eliminated decreases sharply with increasing ADT. AASHTO RRR standards are not as safety cost-effective overall because they do not distinguish between roads with low and high ADT.

The cost for lane and shoulder widening under the 1978 FHWA proposed standards is about 45 percent of the cost under new construction standards. The number of accidents eliminated is about 40 percent, however, so that improvements under the FHWA proposed standards are somewhat less cost-effective than under new construction standards.

Three modifications were made to the 1978 FHWA proposed standards in order to improve their cost-effectiveness. First, the ADT ranges for specific lane and shoulder widths were shifted. The original FHWA proposal defined certain lane and shoulder widths for 1 to 400 ADT, 401 to 4,000 ADT, and

TABLE 5-4 Cost-Effectiveness of Alternative Lane and Shoulder Width Standards, National Level Analysis Using HPMS Data

Land and Shoulder Width Standard	Cost per Year ($millions)[a]	Accidents Eliminated per Year	Cost per Accident Eliminated ($thousands)
AASHTO new construction standards	2,360	58,100	40.6
AASHTO RRR standards	480	8,600	56.0
1978 FHWA proposed standards	1,040	23,800	43.4
FHWA proposed standards with modifications[b]	1,069	33,900	31.5

[a] Costs are in 1985 dollars and were calculated using a 7 percent discount rate and a project life of 30 years.
[b] The 1978 FHWA proposed standards were modified by *(a)* shifting the ADT breakpoints to require more improvements at higher traffic levels and fewer improvements at lower traffic levels, *(b)* increasing minimum shoulder widths for highways with greater than 2,000 ADT levels, and *(c)* reducing minimum shoulder widths for highways on mountainous terrain.

greater than 4,000 ADT. The breakpoint at 400 ADT was shifted to 750 ADT, and the breakpoint at 4,000 ADT was shifted to 2,000 ADT.

The shift from 400 to 750 ADT has the effect of making the standards less stringent—the least stringent lane and shoulder widths (previously applicable to highways with 1 to 400 ADT) are now applicable to highways with 1 to 750 ADT. The shift from 4,000 to 2,000 ADT has the opposite effect—the most stringent lane and shoulder widths (previously applicable to highways with ADT greater than 4,000) are now applicable to highways with ADT greater than 2,000. The net effect of the two shifts is to decrease the cost of lane and shoulder widening, increase the number of accidents eliminated, and decrease the cost per accident eliminated.

The second modification to the 1978 FHWA proposed standards was to increase the shoulder widths required at higher ADT levels from 4 to 6 ft on each side. As noted in the project-level analysis, shoulder widening is more cost-effective than lane widening on most of the nation's highways.

The third modification to the 1978 FHWA proposed standards was to decrease minimum shoulder widths by 1 ft in each direction on mountainous terrain. As shown in the project-level analysis, shoulder widening is less cost effective on mountainous than on flat or rolling terrain.

The combined effect of the three modifications is to increase the cost for lane and shoulder widening by about 3 percent, increase the number of accidents eliminated by about 45 percent, and reduce the cost per accident eliminated by 26 percent. The FHWA proposed standards with modifications could save approximately 1,000 lives and prevent nearly 30,000 injuries each year.[2]

As of January 1986, 26 states had special RRR standards and 24 states used new construction standards for RRR projects. The modified FHWA proposed standards are generally less stringent than special RRR standards at low traffic levels, more stringent than special RRR standards at high traffic levels, and less stringent than new construction standards at all ADT levels.

The cost-effectiveness of standards proposed at the national level were compared with special RRR standards in Florida, Illinois, and Washington state (Table 5-5). In these three states, the modified FHWA proposed standards were about as cost-effective as special RRR standards, but involved higher spending for lane and shoulder widening and eliminated more accidents.

---

[2] These estimates are based on 33,900 accidents eliminated (Table 5-4), distributions of accidents by severity from Smith et al. *(4)*, and information on the number of fatalities per fatal accident and injuries per injury accident from NHTSA *(15)*.

TABLE 5-5 Cost-Effectiveness of Alternative Lane and Shoulder Width Standards: State-Level Analysis for Florida, Illinois, and Washington

State	Lane and Shoulder Width Standard	Cost per Year ($millions)	Accidents Eliminated per Year	Cost per Accident Eliminated ($thousands)
Washington	1978 FHWA proposed standards	13.33	359	37.1
	FHWA proposed standards with modifications	16.43	532	30.9
	State RRR standards	11.13	321	34.7
Florida	1978 FHWA proposed standards	7.70	224	34.4
	FHWA proposed standards with modifications	8.60	364	23.6
	State RRR standards	6.33	316	20.0
Illinois	1978 FHWA proposed standards	10.73	198	54.2
	FHWA proposed standards with modifications	15.89	404	39.3
	State RRR standards	13.62	357	38.2

NOTES: Costs are in 1985 dollars and were calculated using a 7 percent discount rate and a project life of 30 years. The 1978 FHWA proposed standards were modified by (a) shifting the ADT breakpoints to require more improvements at higher traffic levels and fewer improvements at lower traffic levels, (b) increasing minimum shoulder widths for highways with ADT greater than 2,000, and (c) reducing minimum shoulder widths for highways on mountainous terrain.

Summary of Findings on Lane and Shoulder Widening

• Traffic volumes are an important consideration in the cost-effectiveness of lane and shoulder widening. Standards that are not sensitive to traffic volumes, such as the RRR standards presented in the AASHTO RRR guidelines, are not cost-effective.

• Lane and shoulder widening not only eliminate accidents but also result in time savings to highway users. These time savings are relatively small at lower traffic levels but can be an important consideration for highways with ADT greater than 2,000.

• The cost-effectiveness of the lane and shoulder width standards proposed by FHWA in 1978 can be improved by (a) shifting the ADT breakpoints to require more improvements at higher traffic levels and fewer improvements at lower levels, (b) increasing minimum shoulder widths for highways with ADT greater than 2,000, and (c) reducing minimum shoulder widths for highways on mountainous terrain. Modified standards reflecting these changes are given in Table 5-3.

Horizontal Curves

The cost-effectiveness of reconstructing horizontal curves by decreasing their degree of curvature was examined at the project and system levels using *(a)* the safety relationship presented in Chapter 3 and discussed in more detail in Appendix D, *(b)* the cost relationship discussed in Chapter 4, and *(c)* AASHTO methods for estimating highway user travel time and operating cost savings from flattening curves *(13)*.

Standards for horizontal curves are usually expressed in terms of design speeds. As described in AASHTO's *Policy on the Geometric Design of Highways and Streets, 1984 (15)*, the design speed of a curve is determined by radius, superelevation, and the friction generated by the pavement surface.

For new construction, design speed is used as a concept for unifying the various aspects of the new highway. AASHTO defines design speed as "the maximum safe speed that can be maintained over a specified section of highway when conditions are so favorable that the design features of the highway govern" *(15)*. All pertinent features of the highway, including horizontal curves, should be related to the design speed to obtain a balanced design.

The selection of an ideal design speed using a rationale similar to that employed for new construction is generally not practical for RRR projects. It is more practical to compare the as-built design speeds of existing curves with some measure of the speeds of approaching vehicles before they slow down for the curve. The 85th percentile speed is useful for this purpose because it exceeds the speed of most approaching vehicles and has traditionally been used by traffic engineers for setting speed limits.

Project Level Cost-Effectiveness

Important determinants of the cost-effectiveness of flattening horizontal curves are

• Design speed of the curve before and after improvement;
• Other factors influencing accident rate on the curve, such as the distribution of actual operating speeds, whether the curve is isolated on a long tangent section (versus one of many curves on a highway segment), lane width, sight distance, vertical alignment, shoulder width, and roadside environment;
• Cost of flattening the curve, which in turn is affected by the central angle of the curve, cost of additional right-of-way, and unit costs for construction quantities; and

- Traffic levels; the number of accidents eliminated depends not only on the improvement in the accident rate, but also on the volume of traffic using the curve.

Using a hypothetical curve flattening project, the study examined the effect of design speed before and after improvement, central angle, and ADT on cost per accident eliminated. Other factors influencing accident rates are not explicitly accounted for. This is an important limitation of the analysis; because of other factors that must be addressed on a site-specific basis, the accident rate at a curve could be much greater or much less than assumed.

The cost-effectiveness of curve flattening varies sharply with design speed before improvement (Figure 5-7). The cost per accident eliminated for flattening a 30-mph curve is about one-fourth the cost for flattening a 45-mph curve.

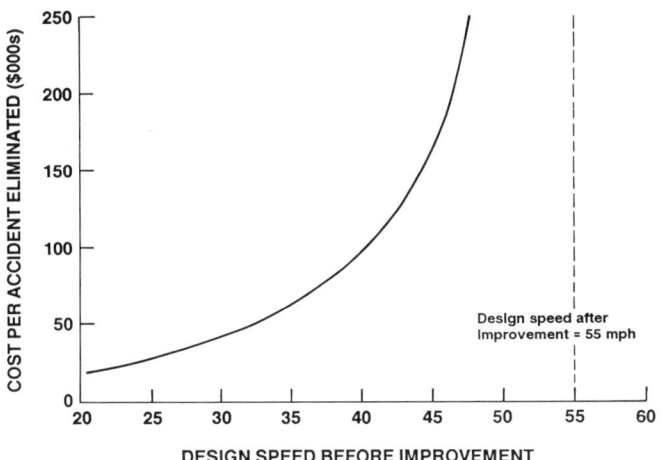

NOTES: Characteristics of the hypothetical horizontal curve on which this sensitivity analysis is based are (a) design speed of 55 mph after improvement, (b) central angle of 30 degrees, and (c) 2,000 ADT. Costs are in 1985 dollars and were calculated using a discount rate of 7 percent and a project life of 30 years.

FIGURE 5-7 Cost-effectiveness of flattening horizontal curves by design speed before improvement.

As discussed in Chapter 4, a relatively large share of the cost for flattening a horizontal curve is fixed and is not sensitive to the amount by which the curve is flattened. For this reason, it is not cost-effective to flatten a curve only slightly because the cost will be out of proportion to the improvement in the accident rate (Figure 5-8). Cost-effectiveness considerations are not just limited to the question of whether or not a given curve should be improved,

150 DESIGNING SAFER ROADS

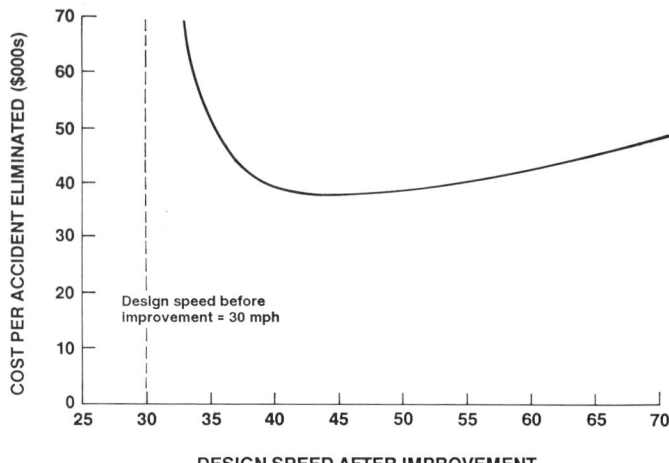

NOTES: Characteristics of the hypothetical horizontal curve on which this sensitivity analysis is based are (a) design speed of 30 mph before improvement, (b) central angle of 30 degrees, and (c) 2,000 ADT. Costs are in 1985 dollars and were calculated using a discount rate of 7 percent and a project life of 30 years.

FIGURE 5-8 Cost-effectiveness of flattening horizontal curves by design speed after improvement.

designers should also consider cost-effectiveness in deciding the degree to which the curve should be upgraded.

The amount of earthwork and right-of-way required to flatten a horizontal curve, and consequently the cost per accident eliminated, is strongly affected by the central angle (Figure 5-9) because the curve length increases roughly in proportion to its central angle.

As with other design features examined in this chapter, cost per accident eliminated varies inversely with ADT (Figure 5-10) because the number of accidents eliminated is directly proportional to ADT.

User savings realized from flattening horizontal curves are appreciable in relation to the cost of these improvements, particularly at ADT greater than 2,000. Taking user savings into account along with safety benefits strengthens the case for flattening curves.

Drivers reduce speed when they approach a curve and then accelerate after they enter or pass the curve. This speed change cycle adds to travel time, fuel consumption, and other components of operating cost. Improving horizontal curves reduces the size of the speed change, which in turn reduces travel times and operating costs for highway users. Curves also affect user costs in two other ways, both of which are insignificant relative to the speed-change effect. First, vehicles expend more energy and experience more tire wear on curves

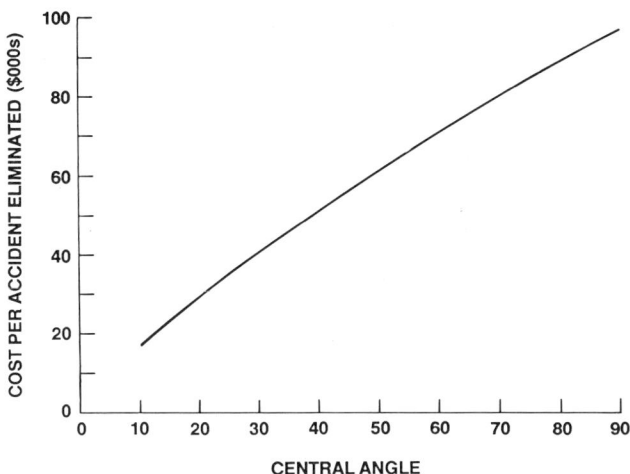

NOTES: Characteristics of the hypothetical horizontal curve on which this sensitivity analysis is based are (a) design speed of 30 mph before improvement, (b) 55 mph after improvement, and (c) 2,000 ADT. Costs are in 1985 dollars and were calculated using a discount rate of 7 percent and a project life of 30 years.

FIGURE 5-9 Cost-effectiveness of flattening horizontal curves by central angle.

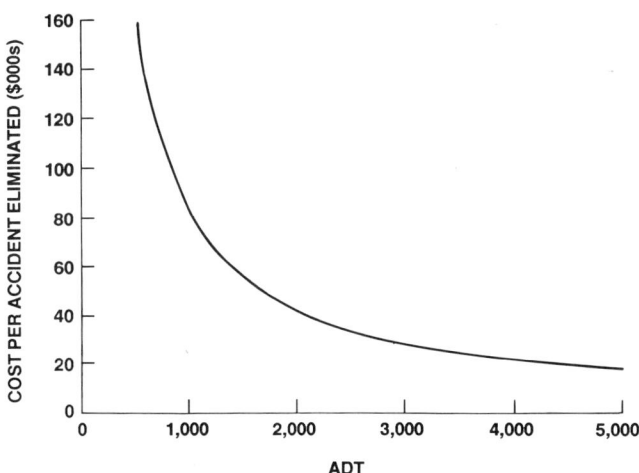

NOTES: Characteristics of the hypothetical horizontal curve on which this sensitivity analysis is based are (a) design speed of 30 mph before improvement, (b) 55 mph after improvement, and (c) central angle of 30 degrees. Costs are in 1985 dollars and were calculated using a discount rate of 7 percent and a project life of 30 years.

FIGURE 5-10 Cost-effectiveness of flattening horizontal curves by ADT.

because of the tire side friction required to keep the vehicle on the curve. Second, flattening a curve provides a small reduction in travel distance.

Using the methodology presented in AASHTO's *Manual on User Benefit Analysis of Highway and Bus-Transit Improvements (12)*, the study estimated the effects of horizontal curves on user costs. The manual provides estimates of the extent to which drivers decrease their operating speeds on curves with low design speeds. The estimates of speed decreases are consistent with more recent estimates *(16)*.

Flattening a horizontal curve to increase its design speed from 30 to 55 mph will reduce user costs by $27 per 1,000 vehicles (Figure 5-11). For a highway with 1,000 ADT, this savings amounts to $10,000 per year.

The $10,000 per year in user savings is a considerable portion of the annualized cost for curve flattening (Table 5-6). At higher ADT levels, user savings for curve flattening might be well in excess of the annualized cost.

System-Level Cost-Effectiveness

The cost-effectiveness of curve flattening was examined by using curve inventory data from state highways in Florida, Illinois, and Washington State.

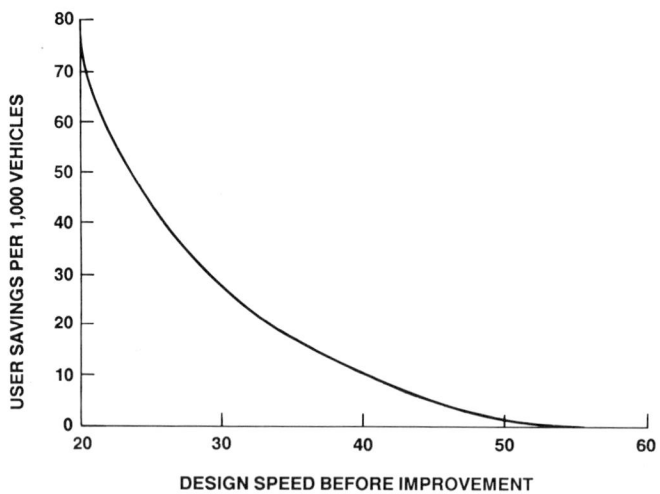

NOTE: Travel time savings are valued at $7.50 per vehicle hour.

FIGURE 5-11 User savings for improving the design speed of a horizontal curve to 55 mph.

TABLE 5-6 Illustrative Cost and User Savings for Horizontal Curve Flattening

Annualized Cost for Curve Flattening	Cost ($)
Added construction cost	13,300
User savings per year	
1,000 ADT	9,900
3,000 ADT	29,700
5,000 ADT	49,500
Net cost (construction cost less user savings)	
1,000 ADT	3,400
3,000 ADT	-16,400
5,000 ADT	-36,200

NOTES: The example assumes design speed before improvement of 30 mph, a design speed after improvement of 55 mph, and a central angle of 30 degrees. Time savings are valued at $7.50 per vehicle hour. Construction cost is annualized using a 7 percent discount rate and a project life of 30 years.

These states maintain detailed roadway inventories that contain the data needed (degree of curve, central angle, and traffic volume) to estimate the cost-effectiveness of curve flattening for individual curves. These data are not available in many other states and are not compiled at the national level.

Curve flattening costs were estimated using the cost relationship from Washington State (Appendix I) with adjustments based on construction price indices for Florida and Illinois. Cost per accident eliminated was calculated with and without accounting for user travel time and cost savings because there is no generally accepted practice for weighting these savings in relation to construction costs.

Although all three states have about the same mileage for two-lane rural highways, Washington State has many more curves with low design speeds reflecting difference in terrain (Figure 5-12). Most of the terrain in Washington is rolling or mountainous whereas most of the terrain in Illinois and Florida is flat.

Cost per accident eliminated for flattening horizontal curves in the three states was estimated by using a design speed of 40 mph as the minimum standard. For highways with 85th percentile speeds of 55 mph or more on tangent sections, under this minimum standard curves will be flattened only when the design speed is more than 15 mph below the 85th percentile speed of vehicles approaching the curve.

Without accounting for user time and cost savings, system-level cost per accident eliminated for flattening horizontal curves with design speeds below the minimum standard of 40 mph ranged from $15,000 in Florida to $100,000 in Illinois (Table 5-7).

Cost per accident eliminated by flattening horizontal curves is higher in Illinois than in Washington and Florida for two reasons. First, most of the

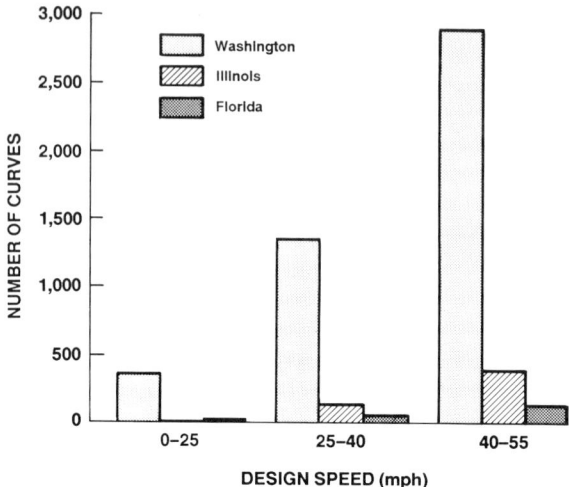

FIGURE 5-12 Number of curves by design speed on two-lane rural state highways.

substandard curves in Illinois are just below the 40 mph minimum, whereas in Washington and Florida, many of the substandard curves are below 30 mph. As shown in the project-level analysis, cost per accident eliminated is very sensitive to design speed before improvement. Second, Illinois has a number of curves with central angles of 90 degrees because roads in Illinois are frequently laid out along the boundaries of rectangular parcels of land. The

TABLE 5-7 Cost-Effectiveness of Flattening Horizontal Curves on Two-Lane Rural State Highways

	Washington	Illinois	Florida
Substandard curves	1,690	145	69
Added construction cost per year	29,271	3,095	1,092
Accidents eliminated per year	481	31	74
Cost per accident eliminated	60.9	99.8	14.8
User time and cost savings per year	28,582	1,745	4,391
Net (construction less user) cost per year	689	1,350	−3,299
Net cost per accident eliminated	1.4	43.5	NM

NOTES: The assumed minimum standard is an as-built design speed of 40 mph. Substandard curves are assumed to be reconstructed with a design speed of 55 mph. When the net cost is negative, net cost per accident elminated is not meaningful (NM). Costs are in thousands of 1985 dollars. Added construction costs are annualized using a 7 percent discount rate and a project life of 30 years.

project-level analysis showed that cost per accident eliminated for reconstructing curves is very sensitive to the central angles of the curves.

Cost per accident eliminated in Florida is much lower than in Illinois and Washington primarily because average traffic volumes are much higher. ADT for the reconstructed curves in Florida is 5,000 versus 1,600 in Washington and 1,700 in Illinois.

When user travel time and operating cost savings are taken into account, curve flattening is substantially more cost-effective. Cost per accident eliminated drops drastically from $61,000 to $1,400 in Washington and from $100,000 to $43,000 in Illinois. In Florida, the user time and cost savings for curve flattening exceed the added cost so that these improvements can be justified at the system level based on user savings alone.

In Washington State, a more detailed stratification of curves by ADT was used to establish the circumstances under which curve flattening is frequently cost-effective (Table 5-8). Taking user savings into account, the average cost per accident eliminated for flattening curves in the 750- to 1,500-ADT range with design speeds greater than 15 mph below 85th percentile speeds is about $20,000. For the 500- to 750-ADT range, however, average cost per accident eliminated is about $100,000, double the $50,000 figure used in this study as an upper limit for improvements justified purely on the basis of safety cost-effectiveness. These findings support the use of 750 ADT and a difference of

TABLE 5-8 Cost-Effectiveness of Flattening Horizontal Curves by ADT: Washington State System-Level Analysis

	ADT				
	Under 500	500–749	750–1,499	1,500 and Over	Total
Substandard curves	246	164	754	526	1,690
Added construction cost per year	4,614	2,982	13,334	8,341	29,271
Accidents eliminated per year	14	18	166	283	481
Cost per accident eliminated	329.6	165.7	80.3	29.5	60.9
User time and cost savings per year	782	1,110	9,855	16,835	28,582
Net (construction less user) cost per year	3,832	1,872	3,479	−8,494	689
Net cost per accident eliminated	212.9	104.0	21.0	NM	1.4

NOTES: The assumed minimum standard is an as-built design speed of 40 mph. Substandard curves are assumed to be reconstructed with a design speed of 55 mph. When the net cost is negative, net cost per accident eliminated is not meaningful (NM). Costs are in thousands of 1985 dollars. Added construction costs are annualized using a 7 percent discount rate and a project life of 30 years.

15 mph between the as-built design speed of curves and 85th percentile speeds as breakpoints to separate situations in which curve flattening is frequently cost-effective from those in which it is not cost effective.

Summary of Findings on Horizontal Curves

• Curve flattening is frequently cost-effective when ADT is greater than 750 and the design speed of the existing curve is more than 15 mph below the 85th percentile speed of vehicles approaching the curve.
• Curve flattening can result in substantial travel time and operating cost savings to highway users. Taking these savings into account strengthens the case for curve flattening.
• A firm, nationwide standard for horizontal curves—such that curves falling below the standard are routinely upgraded—is inappropriate because of the high degree of site-to-site variation in the cost-effectiveness of curve flattening.

Sight Distance at Crest Curves

Standards for sight distance at crest curves are defined in terms of design speed. According to AASHTO's *Policy on the Geometric Design of Highways and Streets (15)*, "the minimum sight distance available on a roadway should be sufficiently long to enable a vehicle traveling at or near the design speed to stop before reaching a stationary object in its path." In calculating sight distance, AASHTO's policy on design assumes the driver's eye is 3.5 ft above the road surface and the object is 6 in. above the roadway. The policy also assumes a brake reaction time (the time interval from the instant the driver recognizes the existence of a hazard on the roadway to the instant the driver actually applies the brakes) of 2.5 sec, which is greater than that required by most drivers under normal highway conditions.

The safety cost-effectiveness of reconstructing crest curves to improve sight distances was examined by using hypothetical projects. Cost per accident eliminated was calculated using the safety relationship presented in Chapter 3 (and discussed in more detail in Appendix E) and the cost relationship discussed in Chapter 4. This analysis probably overstates the cost-effectiveness of flattening a typical crest curve because, as noted in Chapter 3, the safety relationship provides an upper bound to accident reductions resulting from increased sight distance.

Crest curves were not analyzed at the system level because data on cost and accident relationships—length of curve and degree of hazard in the sight-restricted area—are not available.

Project-Level Cost-Effectiveness

Important determinants of the cost-effectiveness of improving sight distance at crest curves are

- Difference between the design speed of the curve and the speeds of vehicles as they travel through the sight-restricted area,
- Degree of hazard in the sight-restricted area, and
- Traffic volumes.

The cost-effectiveness of sight distance improvements at crest curves varies sharply with the difference between design speeds and operating speeds in the sight-restricted area (Figure 5-13). The cost per accident eliminated for improving a curve with a speed difference of 20 mph is about 60 percent of the cost per accident eliminated for improving a curve with a speed difference of 15 mph.

The degree of hazard in the sight-restricted area is also an important consideration. A major hazard (such as a high-volume intersection or sharp horizontal curve) in the sight-restricted area might reduce the cost per accident eliminated by one-half.

As with other design features, cost per accident eliminated varies inversely

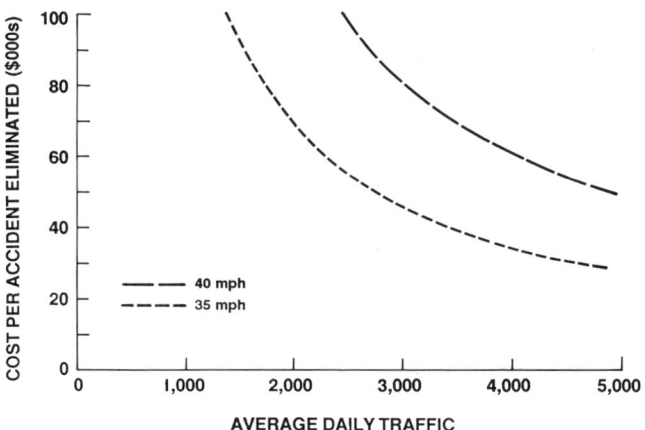

NOTES: Example assumes operating speed of 55 mph and a major hazard in the sight-restricted area. Costs are in 1985 dollars and were calculated using a discount rate of 7 percent and a project life of 30 years.

FIGURE 5-13 Cost-effectiveness of reconstructing crest curves by ADT and design speed before improvement.

158   DESIGNING SAFER ROADS

with ADT because the number of accidents eliminated is directly proportional to ADT.

Lengthening vertical crest curves reduces motor vehicle operating costs by reducing the length of vertical tangents. This effect is small relative to the construction cost of lengthening curves, however.

Consider, as an example, a 600-ft-long vertical curve connecting a 6 percent upgrade and a 6 percent downgrade on a highway with 4,000 ADT. On the basis of estimates of the effect of grades on operating costs in AASHTO's *Manual on User Benefit Analysis of Highway and Bus-Transit Improvements (12)*, lengthening this curve by 1,000 ft will reduce motor vehicle operating costs by $2,600/year. The annualized construction cost for this improvement is about $35,000/year. Thus, even at 4,000 ADT, user savings will cover less than 10 percent of this improvement.

Summary of Findings on Sight Distance at Crest Curves

• Reconstructing crest curves to improve sight distance can be cost-effective when a major hazard (such as a high-volume intersection or narrow bridge) exists in the sight-restricted area, the design speed of the existing curve is more than 20 mph below operating speeds in the sight-restricted area, and ADT is greater than 1,500.
• Highway user travel time and cost savings associated with reconstructing crest curves are generally small relative to the added cost of construction for highways with less than 4,000 ADT.
• A firm, nationwide standard for sight distance at crest curves—such that curves falling below the standard are routinely upgraded—is inappropriate because of the high degree of site-to-site variation in the cost-effectiveness of sight distance improvements.

Bridge Width

The safety cost-effectiveness of bridge width improvements was examined on two-lane rural highways using the relationship between bridge width and accidents presented in Chapter 3 and the relationship between bridge width and cost presented in Chapter 4.

Project-Level Cost-Effectiveness

Important determinants of the cost-effectiveness of bridge width improvements are

- The width of the bridge relative to the width of the travel lanes on bridge approaches;
- Whether widening is practical or, alternatively, whether it is necessary to demolish and reconstruct the bridge to provide more width;
- Whether the bridge requires major rehabilitation or reconstruction for other reasons (e.g., inadequate load-bearing capacity);
- The length of the bridge, which affects the cost of bridge width improvements; and
- Traffic volumes.

The accident rate for bridge-related accidents depends on the relative bridge width—the difference between the width of the bridge and the width of the travel lanes of the highway on which the bridge is located (Chapter 3). If lanes are to be widened as part of a RRR project, the new travel lanes (rather than the travel lanes before improvement) should be used in calculating the relative widths of bridges.

Another important consideration is whether it is practical to widen the existing bridge, or alternatively, whether it is necessary to demolish the existing bridge and construct a new one to provide additional width. Based on the unit costs presented in Chapter 4, widening a 50-ft-long bridge from 20 to 28 ft would cost about $40,000. If widening is not practical, the cost of demolishing the 20-ft-wide bridge and constructing a 28-ft-wide bridge would be about $100,000. Thus, cost per accident eliminated would be more than two times greater if widening is not practical (Figure 5-14).

Bridge widening is less cost-effective for longer bridges. Bridge widening costs increase roughly in proportion to the length of the bridge, but the number of accidents eliminated by widening a narrow bridge is only marginally affected by bridge length. Thus, the cost per accident eliminated is greater for longer bridges.

Bridge width improvements are highly cost-effective when bridges require reconstruction for other reasons (e.g., inadequate load-bearing capacity) because the added cost for providing a wider bridge is substantially less if the bridge is going to be reconstructed anyway.

Designers should consider cost-effectiveness not only in deciding whether to widen a bridge, but also in deciding the extent to which it should be widened. A substantial part of the costs for bridge widening are fixed and must be paid regardless of how much the bridge is widened (Figure 5-15); therefore it is generally not cost-effective to widen bridges by only a few feet. Also, large improvements to bridge width may not be cost-effective because there is little payoff in terms of accident reduction in increasing bridge widths beyond levels mandated by AASHTO new construction standards.

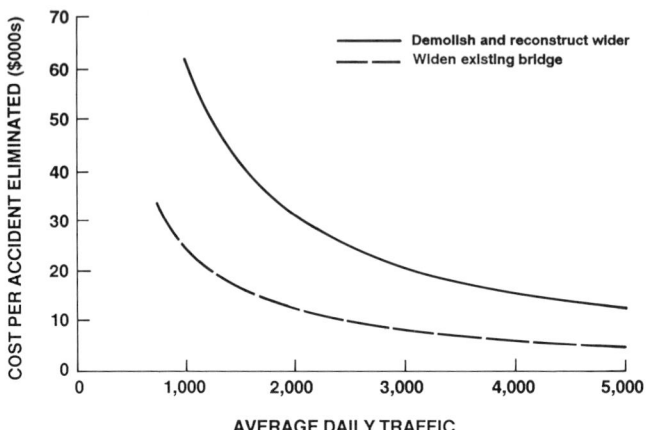

NOTES: Example assumes a bridge length of 50 ft, relative width before improvement of 0, and relative width after improvement of 8 ft. Costs are in 1985 dollars and were calculated using a discount rate of 7 percent and a project life of 30 years.

FIGURE 5-14 Cost per accident eliminated for bridge width improvements.

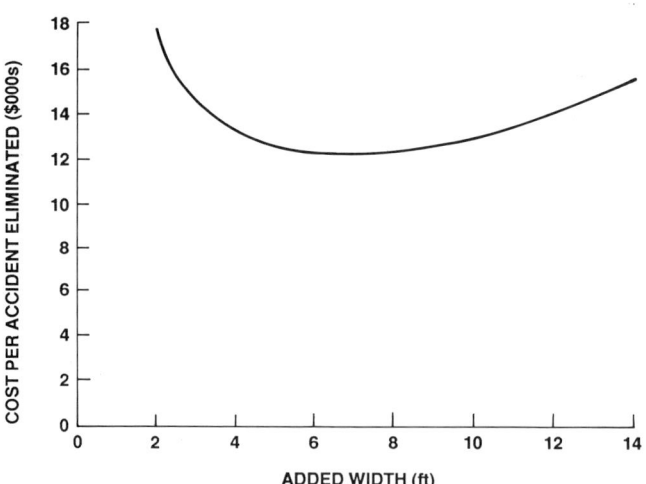

NOTES: Example assumes 2,000 ADT, bridge length of 50 ft, and a relative width of 0 before improvement. Costs are in 1985 dollars and were calculated using a discount rate of 7 percent and a project life of 30 years.

FIGURE 5-15 Effect of added width on cost per accident eliminated for widening an existing bridge.

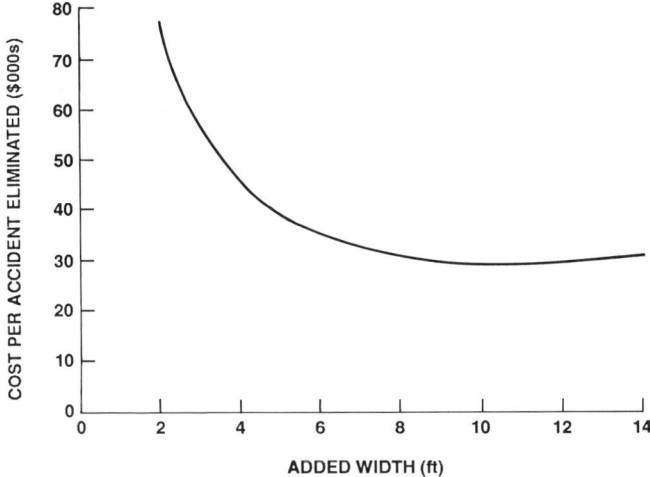

NOTES: Example assumes 2,000 ADT, bridge length of 50 ft, and a width of 0 before improvement. Costs are in 1985 dollars and were calculated using a discount rate of 7 percent and a project life of 30 years.

FIGURE 5-16 Effect of added width on cost per accident eliminated for demolishing a bridge and constructing a wider bridge.

The cost per accident eliminated decreases sharply with added width when an existing bridge is demolished and a wider bridge is constructed in its place (Figure 5-16). Unless there are other considerations in the decision to replace the bridge, such as inadequate load-bearing capacity, this improvement makes sense only for relatively large improvements to bridge width.

User costs are seldom an important consideration in bridge widening on two-lane rural highways. The effect of a narrow bridge on user costs is similar to the effect of a horizontal curve on user costs: drivers reduce their speed when they approach a narrow bridge and accelerate after they cross the bridge. However, field observations of driver behavior at bridges on two-lane rural highways indicate that the amount by which drivers reduce speed is slight (about 2 mph) and not strongly related to the width of the bridge. Instead, drivers compensate for narrow and hazardous bridges by repositioning their vehicles closer to the centerline, even crossing it in extreme cases (17).

A second possible consideration is the role of bridge width as a prime determinant of the bridge's vehicle-carrying capacity. On high-volume highways, a narrow bridge may be a bottleneck at which congestion occurs, which in turn increases travel time and cost. Such congestion seldom occurs on two-lane rural highways.

System-Level Cost-Effectiveness

The cost-effectiveness of bridge width improvements was examined at the system level using data from the FHWA bridge inventory. Cost per accident eliminated was calculated under the assumption that widening the existing bridge is not practical; therefore, it must be demolished and replaced with a wider structure. As suggested by the project-level analysis presented earlier, this assumption will overstate cost per accident eliminated by 50 to 75 percent in those cases in which widening is practical.

This system-level analysis does not address situations in which it is necessary to demolish and replace the bridge for other reasons, such as inadequate load-carrying capacity. In those situations, the cost per accident eliminated for bridge width improvements will be substantially less because the bridge would be demolished in any case and the added cost is only that required to provide a wider structure.

Also, the system-level analysis addressed only bridge width improvements, not lower cost improvements such as improved signing and lanes that gradually narrow as they approach a narrow bridge. The cost-effectiveness of these lower cost improvements could differ greatly from the cost-effectiveness of bridge width improvements.

At the system level, the cost-effectiveness of bridge width improvements varies depending on bridge length, width, and ADT (Table 5-9). For bridges less than 100 ft long with ADT greater than 4,000, the average cost per accident eliminated ranges downward from $30,000 for relative widths less

TABLE 5-9 System-Level Cost-Effectiveness of Bridge Width Improvements

Bridge Length	ADT	Cost per Accident Eliminated by Relative Width		
		0–1.9	2.0–3.9	4.0–5.9
0–100	0–750	224.1	401.6	429.4
	751–2,000	57.6	98.1	137.0
	2,001–4,000	29.2	40.0	59.9
	Over 4,000	15.0	21.9	31.3
Over 100	0–750	803.9	1,619.6	1,875.9
	751–2,000	258.0	420.0	600.5
	2,0001–4,000	162.8	227.9	285.0
	Over 4,000	97.9	139.2	164.7

NOTES: The analysis assumes that existing bridges are demolished and replaced by wider bridges designed to AASHTO new construction standards. The analysis is based on data from the FHWA bridge inventory. Bridges with relative widths less than zero are not included because they are outside the range to which the accident relationship applies. It is likely that cost per accident eliminated for replacing such bridges would be considerably below the values shown in the column for a relative width of 0 to 1.9 ft. Costs are in thousands of 1985 dollars and were calculated using a discount rate of 7 percent and a project life of 30 years.

than 6 ft. For bridges greater than 100 ft long with less than 2,000 ADT, the average cost per accident eliminated is more than $200,000.

Summary of Findings on Bridge Width

• Bridge width improvements are frequently cost-effective when the length of the bridge is less than 100 ft and the width is less than the following values:

Design Year Volume (ADT)	Usable Bridge Width (ft)[a]
0–750	Width of approach lanes
751–2,000	Width of approach lanes plus 2 ft
2,001–4,000	Width of approach lanes plus 4 ft
Over 4,000	Width of approach lanes plus 6 ft

[a]If lane widening is planned as part of the RRR project, the usable bridge width should be compared with the planned width of the approaches after they are widened.

• Highway user travel time and cost savings associated with bridge width improvements are generally small relative to the added cost of making these improvements for highways with less than 4,000 ADT.

• A firm, nationwide standard for bridge width on RRR projects—such that bridges falling below the standard are routinely upgraded—is inappropriate because of the high degree of site-to-site variation in the cost-effectiveness of bridge width improvements.

Roadside Obstacles

The cost-effectiveness of removing roadside obstacles was examined using several hypothetical projects. Cost per accident eliminated was estimated for roadside improvements by using accident relationships presented in Chapter 3 and discussed in more detail in Appendix F. No system-level analyses were conducted because data on roadside obstacles are not compiled by states at the system level.

Project-Level Cost-Effectiveness

Important determinants of the cost-effectiveness of removing a roadside obstacle are

- Degree of hazard posed by the obstacle, for example, as measured by the probability that a collision with the obstacle will result in an injury or fatality;
- Degree of hazard remaining after the obstacle is removed;
- Distance of the obstacle from the edge of the travel lanes;
- Horizontal alignment, as vehicle encroachments on the roadside are more likely to occur at curves; and
- Traffic volumes.

Removal of isolated roadside obstacles can be highly cost-effective, even on low-volume roads. For example, removal of an isolated tree 10 ft from the road can result in a cost per accident eliminated of less than $15,000 at 1,000 ADT and less than $10,000 at 2,000 ADT (Figure 5-17).

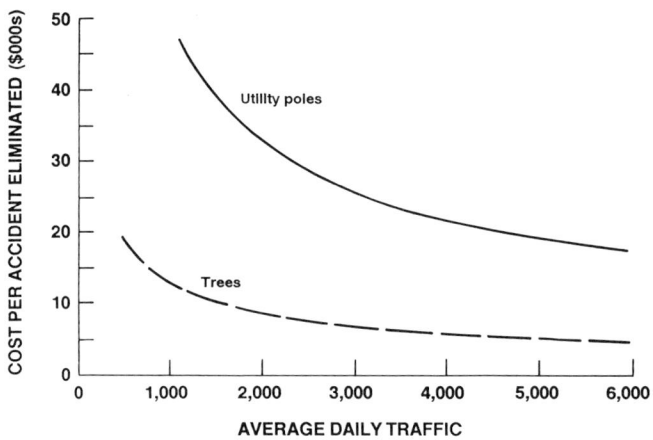

NOTES: Example assumes the obstacle is located 10 ft from the edge of the travel lanes and the area behind the obstacle is a 4:1 fill slope. Costs for removing isolated trees and utility poles are assumed to be $660 and $2,580, respectively (Chapter 4, Table 4-6). Costs are in 1985 dollars and are calculated using a discount rate of 7 percent and a project life of 30 years.

FIGURE 5-17 Cost-effectiveness of removing isolated trees and utility poles by ADT.

In analyzing the cost-effectiveness of removing a roadside obstacle, important considerations include not only the hazard posed by the obstacle, but also the remaining hazards once the obstacle is removed. For example, removing a tree might be highly cost-effective if it is an isolated obstacle on a relatively flat slope. However, if the tree is one of many obstacles or if it is on the edge of a steep slope, then removing it is less cost-effective because drivers who otherwise would have struck the tree are still likely to have a serious accident.

The cost per accident eliminated of removing obstacles increases with distance because there is less chance that errant vehicles will strike obstacles farther from the edge of the road (Figure 5-18). For example, the cost per accident eliminated for removing a roadside hazard located 10 ft from the roadway edge is about 50 percent less than that for removing a hazard located 20 ft from the roadway edge.

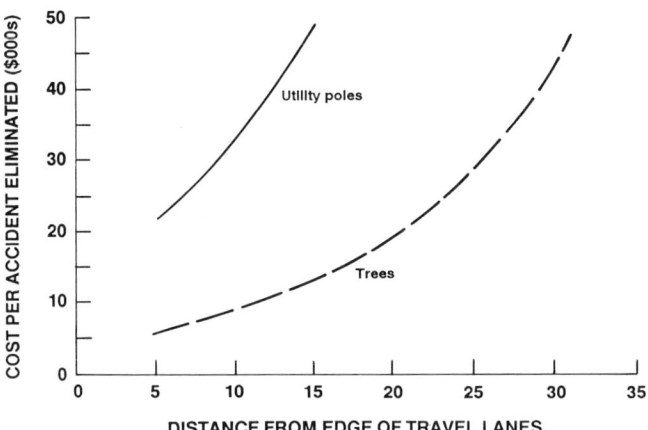

NOTES: Example assumes 2,000 ADT and the area behind the obstacle is a 4:1 fill slope. Costs for removing isolated trees and utility poles are assumed to be $660 and $2,580, respectively (Chapter 4, Table 4-6). Costs are in 1985 dollars and were calculated using a discount rate of 7 percent and a project life of 30 years.

FIGURE 5-18 Cost-effectiveness of removing isolated trees and utility poles by distance from edge of travel lanes.

Although the factors involved in examining the cost-effectiveness of removing roadside obstacles are complex—involving subtleties such as the hazards that remain after an obstacle is removed—it is nonetheless clear that this type of improvement can be highly cost-effective, even on low-volume roads.

Summary of Findings on Roadside Obstacles

• Removal or relocation of isolated roadside obstacles such as trees and utility poles can be highly cost-effective, even on low-volume roads.

- The safety cost-effectiveness of removing a roadside obstacle depends on the distance of the obstacle from the roadway edge, the presence of other obstacles nearby, sideslopes on which the obstacle are located, and traffic volume.

SAFETY-PRESERVATION TRADE-OFFS

Because both pavement resurfacing and geometric improvements for RRR projects are funded from the same source, the trade-off between safety and pavement condition is central to the debate over RRR standards. Changes in design standards for RRR projects could have a substantial effect on pavement condition if the amount of money available for pavement resurfacing is increased or decreased.

About 70 percent of the 540,000 mi of two-lane rural federal-aid highways have lane or shoulder widths less than the widths recommended by AASHTO new construction standards (Table 5-10). The one-time lane and shoulder widening cost to upgrade these highways to new construction standards is about $30 billion. This is about one-half the one-time cost to resurface all 540,000 mi of two-lane rural federal-aid highways. Thus, if new construction standards for lane and shoulder widths were applied to all resurfacing projects, the number of miles that can be resurfaced in a given year would drop sharply without additional funding.

The cost per mile for resurfacing and minor widening of two-lane rural federal-aid highways will vary depending on the width standards used (Table 5-11). Under a fixed budget, about 30 percent more mileage can be completed in a year under AASHTO RRR standards than under AASHTO new construction standards because the per-mile costs for resurfacing and minor widening under AASHTO RRR standards is about 30 percent less. Under the 1978 FHWA proposed standards, and the two variations on these standards,

TABLE 5-10 Deficient Mileage and Widening Costs Under Alternative Standards: National-Level Analysis Using HPMS Data

Alternative Minimum Lane and Shoulder Width Standards	Deficient Mileage (as a Percent of All Two-Lane Rural Federal-Aid Highways)	One-Time Cost for Lane and Shoulder Widening (in billions of 1985 $)
AASHTO new construction standards	71.5	29.3
AASHTO RRR standards	19.7	6.0
FHWA 1978 proposed standards	37.9	12.9
FHWA proposed standards with modifications	35.9	13.3

TABLE 5-11 Cost per Mile Under Alternative Width Standards

Alternative Minimum Lane and Shoulder Width Standard	Resurfacing and Minor Widening Cost per Highway Mile (thousands)
AASHTO new construction standards	174
AASHTO RRR standards	131
FHWA 1978 proposed standards	144
FHWA proposed standards with modifications	145

NOTES: Costs are for all two-lane rural federal-aid highways, as estimated from HPMS data. The cost differences under alternative standards are one-time costs.

about 20 percent more mileage can be completed in a year than under AASHTO new construction standards.

In considering these trade-offs, it should be noted that service lives for geometric design improvements such as lane and shoulder widening are considerably longer than service lives for pavement resurfacing. Once a highway segment is widened to meet standards it does not have to be widened again when the pavement is resurfaced. In contrast to resurfacing, the safety benefits of widening continue over the life of the highway.

Also investigated were the year-by-year safety and pavement preservation consequences of applying alternative lane and shoulder width standards to Washington State's two-lane rural highways over a period of 30 years. Roadway inventory data and pavement deterioration relationships from Washington's Pavement Management System were combined with accident and cost relationships developed in this study. The analysis assumed the following:

• Highway segments will be selected for resurfacing on the basis of pavement condition.
• Projects will be scheduled for improvement until a specified budget is expended.
• Pavements will be improved to new condition if resurfacing is performed; otherwise pavement condition is downgraded to account for deterioration over time.

Four alternatives were examined: *(a)* AASHTO new construction standards, *(b)* modified FHWA standards, *(c)* Washington State RRR standards, and *(d)* resurface only (no lane and shoulder widening). If standards were used, the cross section of the selected project was improved, if necessary, along with resurfacing.

The alternatives varied substantially in their effect on systemwide pavement condition, as measured by the highway mileage with pavements in need

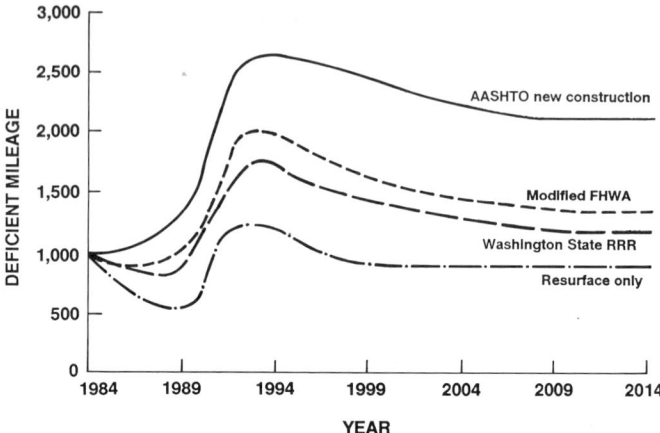

NOTES: Two-lane, rural federal-aid highways only. Pavements are deficient if their condition is below the level where resurfacing must occur according to Washington State pavement management practices (PSI of approximately 2.5). Total system mileage is 5,610.

FIGURE 5-19 Miles of deficient pavement in Washington State under alternative lane and shoulder width standards.

of resurfacing at the end of 1 year (Figure 5-19). All alternatives exhibited a common pattern over time—highway mileage in need of resurfacing peaks about 10 years in the future (because a large number of pavements are due for resurfacing after 10 years), followed by a few years of improvement and eventual stabilization. AASHTO new construction standards have the most adverse effect on pavements: deficient mileage increases sharply and then stabilizes at a level twice that observed in the base year of 1984. Under the Washington State RRR and modified FHWA standards, miles of deficient pavement stabilizes at levels 20 and 40 percent worse than current conditions. Under the resurface-only alternative, miles of deficient pavement stabilizes at a level slightly below that in the base year.

Lane and shoulder widening not only shifts construction funds from current pavement repair, but also influences the process in future years by adding to the surface area that must be maintained and repaved when a segment is resurfaced in the future. If all deficient lanes and shoulders are widened in accordance with Washington State RRR standards, an additional $46 million in systemwide resurfacing costs (3 percent) would result. Under AASHTO new construction standards, systemwide resurfacing costs would increase by nearly $190 million (18 percent) as a result of the extra surface area.

Although stricter standards lead to more deficient pavements, their use also enhances safety (Figure 5-20). Under each of the three sets of standards, the number of accidents eliminated (relative to the resurface-only alternative)

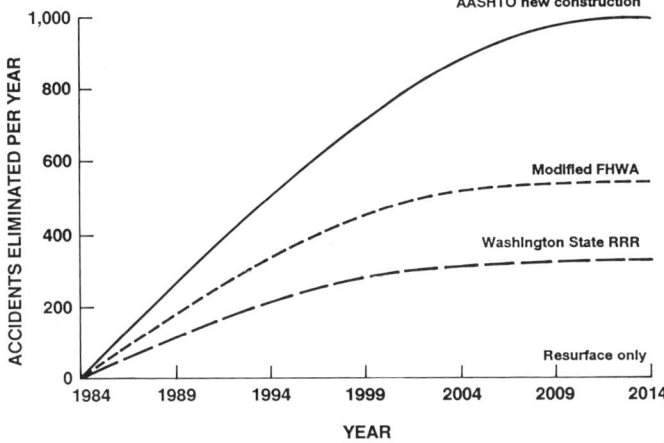

FIGURE 5-20 Accidents eliminated from Washington State highways under alternative lane and shoulder width standards.

increases over time and then stabilizes after most of the segments with deficient lane and shoulder widths have been upgraded. Under AASHTO new construction standards, the number of accidents eliminated per year approaches 1,000. Under less stringent standards, fewer accidents are eliminated—530 per year for modified FHWA and 460 per year for Washington State RRR standards, once all deficient highway segments have been upgraded.

In addition to lane and shoulder widening, other improvements also might involve important safety-preservation trade-offs. The cost of a major effort directed toward curve flattening could be considerable in states that, because of terrain, have many deficient curves. In Washington State, for example, flattening all curves with a design speed less than 40 mph on two-lane rural federal-aid highways would have a one-time cost of $350 million. This is about 40 percent of the cost to resurface all of these highways.

The results from the Washington State analyses illustrate the safety versus pavement preservation trade-off faced by all states: increased spending for safety improvements will reduce the number of miles that can be resurfaced with fixed budgets. However, extrapolating specific results from the analyses to other states must be done with caution because the results depend on existing geometric and pavement conditions. Washington State has far fewer highways with lane widths of 12 ft than do other states, and has somewhat lower shoulder widths than the national average. However, the pavements in Washington State are in better condition than pavements in most states. These factors will determine how many miles can be upgraded on a fixed budget and the safety versus pavement preservation issues faced by each state.

SUMMARY OF FINDINGS

This section contains a summary of general findings on the safety cost-effectiveness of design standards, as well as key findings on specific geometric features.

General

- Traffic levels greatly affect the safety cost-effectiveness of geometric design improvements. Some special RRR standards place little weight on ADT. Fewer accidents and more overall user benefits would result if expenditures for design improvements were shifted to higher volume roads.
- States differ widely on the levels of investment required to achieve a given standard because of factors such as terrain and past design practices. A standard easily achieved in one state might require a considerable portion of the RRR budget in another.
- Changes in geometric design standards for RRR projects can have a substantial effect on pavement condition by increasing or decreasing the amount of money available for pavement resurfacing. For example, the one-time cost of upgrading lane and shoulder widths on all two-lane rural federal-aid highways to AASHTO new construction standards is $30 billion, which is about one-half of the cost to resurface all of these highways.
- Cost per accident eliminated for geometric design improvements made as part of RRR projects frequently falls in the $10,000 to $50,000 range. Improvements with cost per accident eliminated in this range may or may not be warranted depending on the specific value imputed to accidents eliminated; uncertainties surrounding estimates of the added cost for an improvement and the number of accidents it will eliminate; and other factors (e.g., environmental effects) not accounted for in the calculation of cost per accident eliminated.
- Geometric design improvements produced under a given set of design standards will vary in cost-effectiveness because of variations in accident rates, unit costs, design practices, project scale, and other site conditions. Thus, standards must be regarded as tools for identifying situations in which improvements are more likely to be cost-effective. On the one hand, there will

be opportunities for cost-effective improvements beyond those called for by standards. On the other hand, there will sometimes be specific improvements called for by standards that are unusually costly (or that have other undesirable consequences) such that design exceptions are justifiable.

Lane and Shoulder Width

- The cost-effectiveness of the lane and shoulder width standards proposed by FHWA in 1978 can be improved by *(a)* shifting the ADT breakpoints to require more improvements at higher traffic levels and fewer improvements at lower ADT levels, *(b)* increasing minimum shoulder widths for highways with ADT greater than 2,000, and *(c)* reducing minimum shoulder widths for highways on mountainous terrain.
- Highway user travel time savings for widening lanes and shoulders can be significant at higher ADT levels. Taking these savings into account, along with safety benefits, strengthens the case for wider lanes and shoulders on high-volume highways.

Horizontal Curves

- Flattening horizontal curves can be cost-effective, particularly when ADT is greater than 750 and the design speed before improvement is more than 15 mph below the speed of vehicles approaching the curve. However, the considerable site-to-site variation in the cost for alignment improvements indicates that firm nationwide standards for horizontal curves are inappropriate.
- Highway user travel time and operating cost savings for flattening horizontal curves can be considerable. Taking these savings into account, along with safety benefits, strengthens the case for these improvements.

Sight Distance at Crest Curves

- Reconstructing crest curves to improve sight distance can be cost-effective when a major hazard exists in the sight-restricted area, the design speed of the existing curve is more than 20 mph below operating speeds in the sight-restricted area, and ADT is greater than 1,500. However, the considerable site-to-site variation in the cost of these improvements indicates that firm nationwide standards for sight distance at crest curves are inappropriate.
- Sight distance improvements at crest curves provide user time and operating cost savings, but the savings are small in relation to the cost of these improvements and can usually be ignored in cost-effectiveness estimates.

172   DESIGNING SAFER ROADS

Bridge Width

• Bridge width improvements can be cost-effective, particularly when ADT levels are high; the bridge is short (e.g., less than 100 ft); and its width is close to or less than that of the travel lanes on bridge approaches. However, the considerable site-to-site variation in the cost-effectiveness of bridge width improvements indicates that firm nationwide standards for bridge width are inappropriate.
• Bridge widening results in user time and operating cost savings, but the savings are small in relation to the cost of these improvements and can usually be ignored in cost-effectiveness estimates.

Roadside Obstacles

• The removal or relocation of isolated roadside obstacles such as trees and utility poles is frequently cost-effective, even on relatively low-volume roads. However, if an obstacle is close to other hazards, then its removal or relocation is substantially less cost-effective.

REFERENCES

1. Jorgensen Associates. *NCHRP Report 197: Cost and Safety Effectiveness of Highway Design Elements*. TRB, National Research Council, Washington, D.C., 1978, 46 pp.
2. C. V. Zegeer and M. J. Cynecki. "Determination of Cost-Effective Roadway Treatments for Utility Poles." In *Transportation Research Record 970*. TRB, National Research Council, Washington, D.C., 1984, pp. 52–64.
3. J. L. Graham and D. W. Harwood. *NCHRP Report 247: Effectiveness of Clear Recovery Zones*. TRB, National Research Council, Washington, D.C., May 1982, 68 pp.
4. J. A. Smith et al. *Identification, Quantification and Structuring of Two-Lane Rural Highway Safety Problems and Solutions*, Vol. 1. Report FHWA/RD-83/022. FHWA. U.S. Department of Transportation, June 1983.
5. W. F. MacFarland and J. B. Rollins. *Cost-Effectiveness Techniques for Highway Safety: Resource Allocation*. Report FHWA/RD-84/011. FHWA, U.S. Department of Transportation, June 1985.
6. *Guide for Selecting, Locating, and Designing Traffic Barriers*. American Association of State Highway and Transportation Officials, Washington, D.C., 1977.
7. J. C. Glennon. *NCHRP Report 148: Roadside Safety Improvements on Freeways: A Cost-Effectiveness Priority Approach*. TRB, National Research Council, Washington, D.C., 1974, 64 pp.
8. J. R. Leisch and T. R. Neuman. *Study of Width Standards for State Aid Streets and*

Highways. Minnesota Department of Transportation, Report FWHA/MN-79-04. St. Paul, July 1979.
9. *The Economic Cost to Society of Motor Vehicle Accidents.* NHTSA, U.S. Department of Transportation, 1983.
10. *Estimating the Cost of Accidents 1984.* National Safety Council Annual Bulletin, Chicago, Ill.
11. B. C. Kragh, T. R. Miller, and K. A. Reinert. "Accident Costs for Highway Safety Decisionmaking." *Public Roads,* Vol. 50, No. 1, June 1986.
12. *A Manual of User Benefit Analysis of Highway and Bus-Transit Improvements.* American Association of State Highway and Transportation Officials, Washington, D.C., 1977.
13. *Special Report 209: Highway Capacity Manual.* TRB, National Research Council, Washington, D.C., 1985.
14. *Geometric Design Guide for Resurfacing, Restoration, and Rehabilitation (RRR) of Highways and Streets.* American Association of State Highway and Transportation Officials, Washington, D.C., 1977.
15. *A Policy on the Geometric Design of Highways and Streets.* American Association of State Highway and transportation Officials, Washington, D.C., 1984.
16. J. C. Glennon, T. R. Neuman, and J. E. Leisch. *Safety and Operational Considerations for Designs of Rural Highway Curves.* FHWA, U.S. Department of Transportation, Aug. 1983.
17. D. L. Ivey, R. M. Olson, N. E. Walton, G. D. Weaver, and L. W. Furr. *NCHRP Report 203: Safety at Narrow Bridge Sites.* TRB, National Research Council, Washington, D.C., 1979, 63 pp.

6
Tort Liability and Geometric Design

In recent years highway agency administrators have become increasingly concerned about the growth of tort claims. Such claims allege that highway agencies have committed a legal wrong by improper or negligent highway design, operation, or maintenance that became a cause or partial cause of a highway accident. Claims against highway agencies are part of a nationwide problem of rising liability insurance premiums and increasing costs of tort actions.

As a result of tort claim concerns, several important questions have arisen about resurfacing, restoration, and rehabilitation (RRR) standards and design practices:

- Are RRR geometric standards a frequent issue in tort claims?
- If a state adopts and applies special RRR standards less stringent than new construction standards, will this action increase or decrease the chances of subsequent tort claims?
- In what ways can RRR projects affect a highway agency's chances of tort claims?
- What design practices will reduce a highway agency's exposure to successful tort claims?

These questions are explored in light of available statistical evidence, case study findings, and published research. Analysis of information on tort claims in four states (Florida, Louisiana, New York, and Pennsylvania) indicates that the geometric design features covered in RRR standards are usually not the central focus of tort claims. Pavement features, traffic control devices, and roadside barriers account for the large majority of tort claims in Pennsylvania and Louisiana. By correcting hazards related to these conditions as part of RRR projects, highway agencies have an opportunity to reduce the chances of tort claims. Furthermore, by documenting key design decisions on RRR projects, highway agencies improve their defense against tort claims.

BACKGROUND ON TORT LIABILITY

Tort is defined as a civil wrong or injury, and a tort action seeks repayment for damages to property and injuries to an individual. If a defendant is found negligent in his actions or lack of action, he is liable for a tort claim and must compensate the plaintiff. State laws and rulings differ regarding tort claims against a governmental entity. In most states the courts or state legislatures have eliminated sovereign immunity whereby an individual cannot sue the state or its agents for negligence.

Only three states continue to maintain sovereign immunity against tort claims, and even in these states the doctrine has been weakened. A series of court decisions between 1959 and 1961 established the legal foundation for abolishing the doctrine of sovereign immunity *(1)*. In these decisions courts ruled that immunity of states or other government entities was inherently unfair and illogical. Also, past judicial rulings contained numerous exceptions to the doctrine, producing incongruous results. Courts concluded that the states were capable of assuming financial loss from tort judgments, especially with the availability of liability insurance. Increasingly, courts have held government entities responsible for negligent conduct while state legislatures have revised their statutes to allow tort suits against the state for designated activities.

Nevertheless, 47 states have retained a limited form of immunity known as discretionary immunity under which planning and design activities are exempt from liability. Because tort laws vary and local courts rely more heavily on precedents set by their state courts than on those set in other states, guidelines developed for interpreting discretionary immunity and handling tort claims in one state may not be applicable to another.

Concern over the costs of tort claims and liability insurance has grown sharply during the 1980s. An increasing number of claims have received considerable media exposure *(2)* and have led many business and municipal government leaders to advocate some type of tort reform *(3)*. The unavailability of insurance coverage, or large premium increases, have stimulated nationwide discussion of an impending insurance crisis.

The insurance industry contends that the increase in the number and unpredictability of multimillion dollar jury awards has caused large unforeseeable operating losses. Lending support to this, the U.S. Department of Justice Tort Policy Working Group issued a report citing a sevenfold increase in product liability cases handled in federal courts over the past decade *(4)*. A recent survey of 145 cities by the U.S. Conference of Mayors revealed that 15 percent of U.S. cities have canceled or cut back services because of rising tort claims and insurance costs *(5)*. In contrast, a study by the National Center for State Courts termed the so-called "litigation explosion" of the 1980s a myth *(6)*. The study indicated that the increase in total tort claims filed in state courts was no greater than the population increase during the 1980s.

Claims Against Highway Agencies

Highway agencies are spending substantial sums as a result of tort claims. The costs of handling tort claims include not only the direct costs of judgment awards, settlements, and insurance but also attorneys' fees and the cost of engineers' and other staff time.

A series of surveys conducted by the American Association of State Highway and Transportation Officials (AASHTO) provides the best information available on the extent and costs of tort claims against state highway agencies. The latest survey includes information through 1982 *(7)*.

The AASHTO survey indicates that the average amount paid by a state in judgments and settlements during fiscal year 1982 was $894,000, based on 26 states reporting. In some states, however, the costs of tort claim awards and settlements are much greater than these average figures. In fiscal year 1982, for example, California spent more than $5 million on tort claims *(7)*. Pennsylvania spent $72 million on tort claims alone between 1979 and 1986 and currently budgets $25 million per year for awards and settlements *(8)*. From 1982 to 1984 Louisiana was assessed more than $32 million in judgments from tort claims *(9)*. Large cities are also susceptible to large claims—New York City paid more than $6 million in claims in 1985 for pothole-related accidents alone *(10)*.

AASHTO surveys indicate that the average insurance premium for state highway agency liability coverage initially escalated and then declined in the period between 1975 and 1981 *(7)*. The average premium peaked in fiscal year 1976–1977 at about $900,000 per state, but dropped to less than $500,000 by fiscal year 1980–1981. This variation in average premium cost can be attributed to the changes made by state highway agencies in the management of their tort liability exposure and the competition in the insurance industry in the late 1970s and early 1980s. Some states initially increased their insurance coverage, but later dropped coverage completely, became self-insured, and developed in-house expertise to litigate their tort claims. The high interest rates in the late 1970s and early 1980s created favorable investment opportunities for insurance companies causing them to lower rates in an attempt to increase volume.

Basis for Tort Claims Against Highway Agencies

Negligence can be alleged on two grounds particularly relevant to highway agencies:

- Agency (or person) improperly performs its duties (misfeasance).
- Agency (or person) fails to perform its duty (nonfeasance).

State courts use different procedures to establish negligence in an accident. Forty-two states use some form of comparative negligence by which the judge or jury assesses the degree to which each party is responsible for the accident. The initial step in this process is to establish the proximate cause of the accident; that is, the primary cause without which the accident would not have occurred. Proximate cause is often at issue if the plaintiff is in part responsible for the accident; for example, if he was intoxicated and attempts to prove that the roadway feature in question would have caused the accident even if intoxication had not been a factor. Under comparative negligence, if an award is made to the plaintiff each defendant is responsible for paying the assigned percentage of negligence.

However, in approximately 80 percent of the states in which the doctrine of joint and several liability prevails, a defendant's liability is not limited to his percentage of fault—he is potentially liable for all of the award if other defendants cannot pay their share *(11)*. Theoretically, the defendant may bear only 1 percent of the fault but be forced to pay all of the damages when the doctrine of joint and several liability applies. This principle has led plaintiffs to include in their claims defendants with "deep pockets," such as government agencies and large corporations, in order to be assured of receiving a payment if damages are awarded. In response to this action, the legislatures of 11 states have recently passed laws limiting or abolishing joint and several liability *(12–14)*.

Followed in eight states and the District of Columbia, contributory negligence is another procedure courts use to apportion fault. Under this procedure, if the plaintiff is judged at fault in an accident, even if only by a small degree, he is not entitled to claim damages from any other party involved.

As state legislatures or state supreme courts have eliminated or weakened sovereign immunity or adopted comparative negligence as the basis for apportioning fault, opportunities for successful negligence claims against the state have increased. Lawsuits have alleged negligence in virtually every activity of state highway agencies, but maintenance activities are more vulnerable to tort suits *(1)*.

Because maintenance divisions of highway agencies have standard operating procedures, their work can often be easily tested for negligence. Divisions that plan and design roadways, however, must rely on personal judgments, and assessing negligence is not so straightforward. Recognizing this distinction, some states retain discretionary immunity to cover duties of the state that require such judgments, including planning and design activities. Provided a reasonable process is followed, planning and design work may be immune from tort claims.

As states abandoned sovereign immunity, municipalities that had immunity under the umbrella of state law also became vulnerable to tort suits. However, many towns and cities, formed by incorporating, were already susceptible to tort suits because they are considered private corporations.

IMPLICATIONS FOR RRR DESIGN STANDARDS AND PRACTICES

RRR Improvements and Tort Claims

Little is known about how frequently the geometric features addressed by RRR design standards are cited in tort claims against highway agencies. Few states maintain data on tort claims by alleged defect. Further, classifying tort lawsuits is difficult because most involve several defects that differ in importance.

Four states—Florida, Louisiana, New York, and Pennsylvania—maintain summary information on tort claims that indicates alleged defects in roadway condition. As discussed in the following paragraphs, the geometric features addressed by RRR standards are seldom cited in tort claims in these four states. However, the experience of these states may not be representative of other states. In California, for example, state officials indicate that geometric features are more frequently cited in tort cases than experience from the other four states would suggest.

Pennsylvania maintains a summary list of tort claims, beginning at the time the state eliminated its sovereign immunity in 1978, and has compiled information on the number of claims, amount of settlements, and the number of fatalities for the various conditions alleged in tort suits *(8)*. Similar information is available for Louisiana covering the period 1982 to 1984. Louisiana reported the amount of judgments awarded by the court and the number of claims alleging a defect in roadway condition *(9)*. In addition, New York reported the total number of claims by alleged defect in roadway condition for 1983 through 1984 *(15)*, and Florida provided summary data on the number of tort claims filed against the state highway agency by alleged defect in roadway condition *(16)*.

Geometric features usually covered by RRR standards account for a small percentage of the total claims filed against highway agencies in Pennsylvania, Louisiana, New York, and Florida (see Tables 6-1 and 6-2). Data from Pennsylvania and Louisiana are based on awards and settlements, whereas

TABLE 6-1   Costs of Settlements or Judgments in Tort Cases

Condition	Pennsylvania (1979–March 1986)		Louisiana (1982–1984)	
	Amount Paid($)	Percent of Total	Judgments($)	Percent of Total
Geometric features[a]	2,656,094	8.49	401,928	1.24
Pavement[b]	9,434,973	30.16	27,523,066	84.88
Traffic control devices[c]	4,564,075	14.59	3,173,113	9.79
Roadside barriers[d]	5,477,115	17.51	46,266	0.14
Work area protection[e]	1,235,109	3.95	0	0
Other[f]	7,913,942	25.30	1,282,103	3.95
Total[g]	31,281,308[g]	100.00	32,426,476	100.00

[a] Cross section, alignment, intersections.
[b] Pavement edge drops, slippery pavements, potholes, surface deterioration.
[c] Signs, signals, pavement markings.
[d] Guardrails, median barriers.
[e] Temporary conditions during construction.
[f] Employee operations, weather-related conditions, other roadway features.
[g] Claims filed after Pennsylvania's Sovereign Immunity Act of September 1978. The total paid in tort claims was $72 million, which includes the listed claims, plus approximately 450 claims submitted before September 1978 and claims that were not coded with the listed claims. Cases filed after September 1978 – the date of the act – are subject to a payment limit of $250,000 per person and $1 million per accident.

data from New York and Florida are based on tort claims filed against the highway agency. Cross section, alignment, and intersections account for 8.5 percent of costs in Pennsylvania and 1.2 percent of judgment costs in Louisiana. Of the cases in which a geometric feature is at issue, horizontal and vertical curves (4 percent of total costs in Pennsylvania and 1.2 percent of judgment costs in Louisiana) are the most often cited. The importance of geometric features in tort claims may have been obscured in Louisiana because of the unusually large percentage of claims related to pavement condition.

Improper design of geometric features represents 8.0 percent of total tort claims filed in New York from 1983 to 1985. Data from Florida indicate generally similar results; between July 1972 and November 1986 design of geometric features accounted for 7.1 percent of total claims filed.

TABLE 6-2 Tort Claims Filed in New York and Florida

Condition	New York (1983–1985)		Florida (1972–1986)	
	No. of Claims	Percent of Total	No. of Claims	Percent of Total
Geometric features[a]	467	8.0	556	7.1
Pavement[b]	1,016	17.4	2,019	25.8
Traffic control devices[c]	1,238	21.2	876	11.2
Roadside barriers[d]	397	6.8		
Work area protection[e]	426	7.3	4,374	55.9
Other[f]	2,295	39.3		
Total	5,839	100.00	7,825	100.00

[a] Cross section, alignment, intersections.
[b] Pavement edge drops, slippery pavements, potholes, surface deterioration.
[c] Signs, signals, pavement markings.
[d] Guardrails, median barriers.
[e] Temporary conditions during construction.
[f] Employee operations, weather-related conditions, other roadway features.

In Pennsylvania the Office of Attorney General settles most tort claims out of court, whereas in Louisiana tort claims are more often decided by a judge.[1] Neither Pennsylvania nor Louisiana reports cross-section features, which include lane widths, shoulders, sideslopes, and ditches, as the basis for any tort claims. Curve alignment accounts for nearly 50 percent of the tort claim settlements in the geometric category in Pennsylvania. Intersections account for a smaller percent of total settlements. In 1982 five claims in Louisiana alleged inadequate sight distance.

Pavement features, including edge drops, potholes, surface deterioration, and slippery pavements, account for nearly one-third of the settlement costs in Pennsylvania and more than 80 percent of the judgment costs in Louisiana. Of these features, pavement edge drops account for more than 20 percent of settlement costs in Pennsylvania, whereas shoulder condition (which includes edge drops) accounts for 66 percent in Louisiana.

Although geometric features account for slightly less than 2 percent of the total number of claims in Pennsylvania, they account for more than 8 percent of the associated fatalities.

---

[1] Louisiana laws on tort liability are unique in that they are based on the Napoleonic Code of Civil Law, whereas the laws of other states are based on English Common Law.

Susceptibility of RRR Projects and Standards to Tort Claims

The standards selected for RRR projects, the design process followed, and the scope of the improvements may influence the litigation of future tort claims. The issues that might arise in a tort action are

- Did the project meet the appropriate design standards?
- Are the standards reasonable?
- Was the design process reasonable?
- Did the improvements correct existing dangers?
- Should unimproved roads be judged by standards used for roads that are improved?

The resolution of tort claims alleging an inadequate geometric design is contingent on determining the appropriate set of design standards used to assess negligence. New construction design standards are continuously updated. In most states, the standards in effect at the time a roadway was constructed or reconstructed are generally used to assess the adequacy of the road. For example, in the 1972 case of *Hampton v. State Highway Commission, Kansas*, the court held that liability could not be predicated on design defects alone because the design was adequate when the highway was built and must be judged by standards prevailing at the time *(1)*.

Determining whether a highway improvement project is sufficiently extensive to qualify as reconstruction can be a key issue in a tort claim because reconstruction projects usually must meet current new construction standards. For example, the definition of reconstruction was central to the 1985 case of *Brend v. Iowa* in which the judge ruled that resurfacing and seal coating do not constitute reconstruction and therefore the standards of the period during which a highway was originally built continue to apply *(17)*.

Even when a highway has not been reconstructed, the adequacy of the existing design may be challenged when conditions have changed substantially such that a clear danger exists. For example, the state of California lost a tort case *(Baldwin v. State)* in which the omission of left-turn lanes was known to be dangerous. Although the highway was designed in accordance with applicable standards, traffic conditions had changed over time. The court ruled that "once the entity has notice that the plan or design, under changed physical conditions, has produced a dangerous condition of public property, it must act reasonably to correct or alleviate the hazard" *(1)*.

The RRR project design process provides an opportunity for highway agencies to review safety and traffic conditions. An accident analysis and site inspection should reveal any hazardous conditions, which, if left unattended, could become the basis of a future tort action.

Deficient roadside signs or pavement markings and pavement edge-drop problems, which are often the bases of tort claims, can be routinely corrected on RRR projects.

Defense of a RRR Project Design

Although planning and design activities are exempt from liability in most states, this immunity has been held not to apply to decisions made without prior study or conscious deliberation. In *King v. State*, for example, the court held that discretionary immunity was not available to a state highway agency that failed to exercise "due care" in planning a traffic light system *(1)*. In such cases, documentation of the planning process should be a part of the state highway agency's defense.

For RRR projects, documentation should demonstrate that safety aspects of the roadway design were properly considered. Reports that identify deficiencies in existing roadways are potentially threatening to the public agency preparing the report if the deficiencies are not addressed. Thus, if an exception to an applicable design standard was granted, documentation should explain the reasons for the exception and show that logical and orderly procedures were followed in obtaining it.

When a highway agency contemplates a design exception for a geometric or roadside feature, it should be prepared to prove why the feature need not meet the same standards as other facets of the roadway design. Often, the best defense in this situation is to demonstrate that the safety cost-effectiveness of further upgrading the feature does not meet any reasonable criteria. Part of this defense is evidence that special care was taken in determining that an exception was appropriate.

Courts seldom rule that the unavailability of funds is justification for not correcting an alleged defect, but the issue of availability of funds can be part of the defense in relation to the agency's programming procedures. The following points are important to such a defense: the agency is aware of the condition of its facilities; deficiencies have been ranked on a logical basis; and, given the existing funding, are being corrected in the order of priority. Appropriate warnings or other temporary measures should be used to alert the public that deficiencies have not been corrected *(18)*; the highway agency can then affirm that it has performed its duties in the best way possible with the available resources. However, in some states such as Pennsylvania, courts allow neither inadequate funding nor more critical safety priorities elsewhere as a justification for inaction.

Sometimes segments of a highway may be designed to new construction standards and other segments may simply be upgraded according to less stringent RRR standards. A highway agency involved in a tort case under such circumstances should use a defense similar to the one it offers for a design exception, and, again, costs of construction alone are not a sufficient criterion. If the process by which portions of a roadway are selected for rehabilitation instead of reconstruction is not arbitrary, courts will usually hold that RRR standards are applicable.

If RRR design standards are directly implicated in a tort case (e.g., the standards are alleged to be too lenient), the state highway agency may have immunity from liability. As noted previously, many state legislatures have granted states immunity from liability for planning and design activities, which include setting standards. Such activities are considered discretionary and require the special expertise of trained personnel. States without sovereign immunity are clearly liable for operational activities such as day-to-day maintenance. As stated in the landmark case *Weiss v. Fote* (New York, 1960), "Lawfully authorized planning by governmental bodies has a unique character deserving of special treatment as regards the extent to which it may give rise to tort liability. To accept a jury's verdict as to the reasonableness and safety of a plan of governmental services and prefer it over the judgment of the governmental body which originally considered and passed on the matter would be to obstruct normal governmental operations and to place in inexpert hands what the legislature has seen fit to entrust to experts" *(1)*.

In order to receive immunity for planning and design activities, a state must thoroughly document the design process in order to defend challenges. The ruling in *Weiss v. Fote* set certain conditions under which states should not be granted immunity from liability for planning and design activities: *(a)* the plan or design was not duly considered, *(b)* there is no evidence that due care was used in preparation of the plan or design, *(c)* no reasonable official could have accepted the plan, and *(d)* approval of the plan was arbitrary. Simply stated, a rational and orderly process must be followed if a plan or design is to be considered immune from claims of negligence. If a feature built during construction was not called for in the plans or was altered from the specifications, it is open to a claim of negligence in a tort action.

SUMMARY

Available data on the characteristics of tort claims against highway agencies indicate the following:

- Geometric features account for a small percentage (about 10 percent or less) of tort claims in the four states with available data. The experience of these states may be too atypical to generalize, but it is probable that geometric features account for less than one-fourth of all tort claims.

- RRR projects routinely correct or upgrade features such as pavement edge drops, signing, guardrails, and median barriers, often the targets of tort claims.
- When roadway geometrics are an issue in a tort claim related to a RRR project, application of RRR design standards, less stringent than new construction standards, is unlikely to be a basis for the tort claim. Most states have some type of design immunity that may cover the use of design standards as long as reasonable procedures are followed.

Highway agencies can minimize their susceptibility to tort claims by

- Using the RRR design process as an opportunity to identify and correct hazardous conditions, with special emphasis on nongeometric features prone to tort actions;
- Documenting the reasons for making a RRR improvement on a highway segment rather than reconstructing it, particularly if nearby segments are being reconstructed; and
- Documenting the entire design process with particular emphasis on the rationale for seeking design exceptions.

REFERENCES

1. L. W. Thomas, ed. *Selected Studies in Highway Law*. Vol. IV, TRB, National Research Council, Washington, D.C., 1982.
2. "Courting Disaster: Lawsuit Costs Keep Rising, Affecting Economy. *The Wall Street Journal*, May 16, 1986.
3. "Finger-Pointing Distinguishes Attempts to Fix Blame for Liability Crisis." *National Journal*, Feb. 15, 1986.
4. *Report of the Tort Policy Working Group on the Causes, Extent, and Policy Implications of the Current Crisis in Insurance Availability and Affordability*. U.S. Department of Justice, Feb. 1986.
5. *Municipal Liability Concerns: A 145-City Survey*. U.S. Conference of Mayors, Washington, D.C., July 1986.
6. R. T. Roper. *The State Court Case Load Statistics, 1984 Annual Report*. National Center for State Courts, Williamsburg, Va., 1986.
7. *Survey on the Status of Sovereign Immunity in the States, 1983*. Administrative Subcommittee on Legal Affairs, American Association of State Highway and Transportation Officials, Washington, D.C.
8. Unpublished data. Pennsylvania Department of Transportation, Harrisburg, March 1986.
9. Unpublished data. Louisiana Department of Transportation and Development, Baton Rouge, Jan. 1986.
10. "New (Bonding) Agent in the War on Potholes." *The New York Times*, Nov. 12, 1986.

11. *The Need for Legislative Reform of the Tort System.* American Tort Reform Association, Washington, D.C., 1986.
12. "Insurance Woes Spur Many States to Amend Law on Liability Suits." *The New York Times*, March 31, 1986.
13. "Limits on Lawsuit Awards Are Voted in California." *The Wall Street Journal*, June 5, 1986.
14. "Insurance Firms Profit From Crisis." *The Washington Post*, Dec. 21, 1986.
15. Unpublished data. New York Department of Transportation, Albany, Jan. 1987.
16. Unpublished data. Florida Department of Transportation, Tallahassee, Nov. 1986.
17. "State Wins: Road Was Not 'Reconstructed'." *TranSafety Reporter*, Aug. 1985.
18. *NCHRP Synthesis of Highway Practice 106: Practical Guidelines for Minimizing Tort Liability.* TRB, National Research Council, Washington, D.C., 1983, 40 pp.

7
Findings and Recommended Design Practices for Resurfacing, Restoration, and Rehabilitation Projects

Summarized in this final chapter are findings on the selection and design of resurfacing, restoration, and rehabilitation (RRR) projects; the cost and safety trade-offs involved in improving geometric features on these projects; and the influence of design standards. Recommended practices are presented that will promote more safety-conscious design and, in turn, more safety cost-effective RRR projects.

FINDINGS

Resurfacing, restoration, and rehabilitation projects enable highway agencies to improve highway safety by selectively upgrading existing highway and roadside features without the cost of full reconstruction. For example, widening lanes and shoulders on two-lane rural highways on the federal-aid systems alone could save about 1,000 lives and prevent nearly 30,000 injuries each year (Chapter 5). In the last few years many highway agencies have paid increasing attention to safety on federal-aid RRR projects. Nevertheless, further opportunities for safety cost-effective improvements on RRR projects often exist. Highway agencies can take additional steps to improve highway safety on RRR projects while performing pavement repairs or other highway preservation activities.

In practice, however, most highway agencies have had difficulty in striking a balance between these two objectives. This happens in part because the effect of RRR projects on highway service life is relatively predictable whereas the effect on safety appears less certain. Inadequate information about the safety payoff of improvements to existing highways underlies much of the confusion and difference of opinion about the appropriate level of safety-motivated improvements on RRR projects.

Although relationships between safety and highway geometry are not clearly understood, available data demonstrate that highway design strongly influences highway safety. Moreover, for selected features on two-lane rural highways, the study committee found sufficient data and research to make judgments about the most probable relationships so that it could analyze the safety and cost trade-offs involved in improving these features. These analyses show that improving existing highways to match new construction design standards is generally unwarranted. By permitting more highway miles to be improved earlier, less stringent RRR design standards can better enhance systemwide safety.

The review of current RRR design practices conducted for this study concluded that federal-aid RRR projects usually do enhance safety and that highway agencies generally have paid greater attention to safety since Congress added safety enhancement as a RRR objective. But the review, coupled with the study's cost-effectiveness analyses, also concluded that many opportunities for low-cost safety improvements are neglected and that RRR funds currently spent for safety improvements could be redirected for greater systemwide safety gains. A number of factors are responsible for this situation:

- *RRR design practices vary widely from agency to agency.* Some highway agencies follow exemplary practices to address safety needs, some of which are incorporated in later recommendations. Others do not place enough emphasis on evaluation of existing conditions and accident histories to detect safety needs and analysis of opportunities for meeting these needs.

- *RRR projects are initiated primarily to address pavement repair and rehabilitation needs.* Safety needs are often not addressed until a project has been programmed and preliminary design has been initiated. By then, little schedule flexibility remains to accommodate geometric improvements that require additional time for design or right-of-way acquisition.

- *Federal-aid RRR projects frequently widen lanes and shoulders but seldom reconstruct sharp curves or replace bridges with narrow decks.* Because there is a higher concentration of accidents at curves and bridges, improvements at these locations, despite the high costs, can sometimes be more cost-effective with respect to safety than routine cross-section improvements.

- *Not enough is known about the safety gains that will occur after the geometry of existing highways is improved or other safety-oriented improvements are made.* Available information is not always in the hands of designers, or in a form that can be applied without ambiguity.

- *Engineers who administer state traffic and safety programs seldom participate in the design of RRR projects.* They are usually the agency staff members most knowledgeable about accident data and special safety measures, but they have other duties and assignments. RRR project designers often have not received sufficient education in safety engineering.

Design standards alone cannot address these factors that collectively limit the safety gains of federally funded RRR projects. Within the overall process of planning, selecting, and designing RRR projects, the influence of safety standards is small. RRR standards, which can affect only a few key design features, cannot be tailored to fit all possible, or even most, circumstances and constraints encountered in a given state or a specific site. Both the special RRR standards proposed by the Federal Highway Administration (FHWA) and the American Association of State Highway and Transportation Officials (AASHTO) for nationwide use on federal-aid RRR projects and those adopted for use in individual states seldom contain firm minimum standards for features other than lane and shoulder widths.

Furthermore, where firm lane and shoulder width standards are applicable for federal-aid projects, such standards can be circumvented. For instance, some highway agencies reserve federal aid for pavement repairs on projects in which lane and shoulder widths already meet the standards. Pavement repairs on highways for which cross-section improvements are most needed are deferred or undertaken using state or local funds so that widening either is not required or is determined by less stringent standards.

A variety of practices are recommended that encompass the entire RRR process but with special focus on design practice. In selected instances, federal, state, and local highway agencies can use the recommendations, along with design aids, published manuals, and local experience to develop or modify minimum design standards for RRR projects. The Secretary of Transportation is required by statue to ensure that, for federal-aid RRR work, projects are designed and constructed in accordance with standards that extend the service life of highways and enhance highway safety. To accomplish this, the Secretary, acting through the Federal Highway Administration, must either set nationwide RRR standards or approve standards adopted by individual states. For either approach, the committee's recommendations provide guidance.

The recommended design practices provide for a safety impact evaluation in which the safety consequences of existing conditions and potential improvements are evaluated to develop more safety-conscious designs. This process should improve the procedures for selecting the most safety cost-effective improvements.

If these recommendations are followed for federal-aid projects on nonfreeway highways, project spending for lane and shoulder widening will generally

TABLE 7-1 Organization of Study Recommendations

Safety-Conscious Design Process
  1. Assessment of Site Conditions Affecting Safety
  2. Determination of Project Scope
  3. Documentation of the Design Process
  4. Review by Traffic and Safety Engineers

Design Practices for Key Highway Features
  5. Minimum Lane and Shoulder Widths
  6, 7. Horizontal Curvature and Superelevation
  8. Vertical Curvature and Stopping Sight Distance
  9. Bridge Width
  10. Sideslopes and Clear Zones
  11. Pavement Edge Drop and Shoulder Type
  12. Intersections
  13. Normal Pavement Crown

Other Design Procedures and Assumptions
  14. Traffic Volume Estimates for Evaluating Geometric Improvements
  15. Speed Estimates for Evaluating Geometric Improvements
  16. Design Values for Geometric Improvements
  17. Design Exceptions

Planning and Programming RRR Projects
  18. Screening of Highways Programmed for RRR Projects
  19. Assessment of the Systemwide Potential for Improving Safety

Safety Research and Training
  20. Special Task Force to Assess Highway Safety Needs and Priorities
  21. Compendium of Information on Safety Effects of Design Improvements
  22. Increased Research on the Relationships Between Safety and Design
  23. Safety Training Activities for Design Engineers

decline and spending for alignment, bridge, roadside, and intersection improvements, as well as project design, should increase. In some states these shifts may decrease typical RRR project costs; in others they will increase typical costs. Nationwide, the typical project cost will probably increase slightly but not enough to measurably interfere with RRR pavement repair and preservation activities.

Although the study provides guidance in many areas (such as the relationships between accidents and specific geometric features), more detailed guidelines, further research, and increased safety education for designers will be needed to address many of the questions that remain about how to best enhance safety through highway design and operations. Nevertheless, the study committee concluded that a design process that emphasizes safety, even with some parts that will be handled differently by different agencies and designers, will ultimately produce better RRR projects.

The 23 study recommendations are organized into five categories (Table 7-1):

1. Safety conscious design process—recommended general practices that comprise a systematic process for considering safety during RR project design.

2. Design practices for key highway features—recommendations that specify the existing highway conditions that warrant a geometric improvement outright or a serious evaluation of a geometric improvement, as well as design practices that should be followed routinely for key features.

3. Other design procedures and assumptions—recommendations that cover design factors such as speed and traffic volume, as well as guides for selecting design values for geometric improvements and seeking design exceptions.

4. Planning and programming RRR projects—recommendations for including safety considerations in the planning and programming of RRR projects.

5. Safety research and training—recommendations for increased attention to highway safety research, training, and education.

These recommendations generally apply to nonfreeway RRR projects whether or not they are funded with federal aid. For federal-aid RRR projects in particular, the Secretary of Transportation, through the Federal Highway Administration, should take the necessary steps to implement the recommendations in the first three categories—safety-conscious design process, design practices for key highway features, and other design procedures and assumptions. Taken together, these recommendations (Recommendations 1 through 17) comprise a practical national policy on RRR project design that will be more safety cost-effective and comprehensive than an extensive set of rigid minimum standards.

The fourth category of recommendations, planning and programming RRR projects, is directed to state and local highway agencies that have the authority to perform these functions for federal-aid projects without federal oversight. The final category, safety research and training, is directed to the larger highway community with specific recommendations intended for the Congress, FHWA, AASHTO, and state and local highway agencies.

SAFETY-CONSCIOUS DESIGN PROCESS

Significant improvements in safety are not automatic by-products of RRR projects; safety must be systematically engineered into each project. To do this, highway designers must deliberately seek safety opportunities specific to each project and apply sound safety and traffic engineering principles. Highway agencies must strengthen safety considerations at each major step in the design process, treating safety as an integral part of design and not as a secondary objective. These actions require that highway agencies devote greater resources to RRR project design, and to be successful, the added design labor and effort must be complemented with safety-oriented training.

Assess Current Conditions

Recommendation 1: At the beginning of RRR project design, highway designers should assess existing physical and operational conditions affecting safety:

- *Analyze accident and travel data* to identify specific safety problems that might be corrected and to determine if the site has been unduly hazardous compared with the systemwide performance of similar highways.
- *Conduct a thorough site inspection* using personnel trained to identify features that pose safety hazards under common operating conditions and recognize opportunities for safety improvements.
- *Determine and verify existing geometry,* including lane and shoulder widths; degree, length, and superelevation of horizontal curves; stopping-sight-distance restrictions; locations and design of intersections; sideslopes; and clear recovery distances.
- *Determine prevailing speeds* at approaches to horizontal curves and at curves or hill crests with possible stopping-sight-distance restrictions.

The designers of RRR projects can draw on substantial information that bears directly and indirectly on safety. Unlike designers of new highways, designers of RRR projects work with a highway in operation, one with an established safety record, and one for which design and operational characteristics can be observed and measured. Not all state highway agencies take advantages of these favorable circumstances; few perform all of the activities recommended. Many failures by highway agencies to correct safety hazards or to make low-cost safety improvements on RRR projects can be traced to a breakdown in the design process at this early stage.

Determine Project Scope

Recommendation 2: In addition to pavement repairs and geometric improvements, designers of RRR projects should consider and, where appropriate, incorporate other intersection, roadside, and traffic control improvements that may enhance safety.

Such improvements should be routinely considered by designers and should include improvements at intersections and driveway entrances to increase sight distance and reduce vehicle conflicts; replacement or rehabilitation of obsolete bridge rails and guardrails; removal of roadside obstacles and unnecessary guardrails; slope flattening; ditch relocation and regrading; and new or improved signing, pavement markings, and other traffic control devices.

192   DESIGNING SAFER ROADS

These improvements can provide significant reductions in the frequency and severity of accidents. The safety benefits of these improvements, coupled with their low costs, are such that they can be highly cost-effective on RRR projects. In its review of completed federal-aid RRR projects, the FHWA concluded that many highway agencies missed opportunities to enhance safety, often at low cost, through improvements such as those described (Chapter 2). Such omissions occur when designers conceive their missions too narrowly, for example, concentrating on pavement repairs or confining the safety review to a few key highway features.

Document the Design Process

Recommendation 3: Before developing construction plans and specifications, designers should prepare a safety and design report. Safety components of this report should describe the following:

- *Existing geometric and roadside features, traffic volumes and speeds, and accident history;*
- *Applicable minimum design standards;*
- *Specific safety problems or concerns raised by accident data, field inspection, or concerns expressed by the public;*
- *Design options for correcting safety problems and the cost, safety, and other relevant impacts of these options;*
- *Proposed exceptions to applicable design standards and the rationale to support the exceptions; and*
- *Recommended design proposal and its cost and safety impact.*

Approximately one-third of the state highway agencies reviewed for this study currently use some form of design report to document the design process (Chapter 2). Documentation of the design process improves design decisions and generally increases the accountability of those involved in the design, allowing more meaningful design reviews. Moreover, good documentation is a sound defense against tort claims alleging improper design (Chapter 6).

The length and complexity of the recommended safety and design report will vary depending on existing conditions and the extent of any safety problems.

Review the Design

Recommendation 4: Traffic and safety engineers should routinely review safety and design reports, as well as proposed RRR designs before final approval.

Direct participation by traffic and safety engineers in the design of RRR projects contributes to a more safety-conscious design process. However, such participation is often not feasible. Design divisions or units within state highway agencies are generally responsible for the design of RRR projects. Because traffic and safety programs are often administered elsewhere (frequently in maintenance divisions), traffic and safety specialists rarely participate in the design of RRR projects. They have other responsibilities and are not always based in district or regional offices where they can work directly with design teams.

Nevertheless, it is feasible to have traffic and safety specialists routinely review and critique RRR design reports. Such a critique could be accomplished as part of design reviews common in many state highway agencies and could be coordinated with training activities aimed at improving the safety engineering skills of designers (see Recommendation 23).

DESIGN PRACTICES FOR KEY HIGHWAY FEATURES

Recommended design practices include minimum values that can be used by FHWA and state and local highway agencies in setting minimum RRR geometric design standards for selected features. Designers use such standards to determine whether a particular geometric feature must be upgraded as part of a RRR project. A feature not meeting the minimum standard must be upgraded unless a design exception is sought and approved.

Numerical minimum RRR standards are warranted for nationwide use when the following conditions are met:

- Trade-offs between safety and performance against cost can be evaluated quantitatively, and conclusions can be drawn about the safety cost-effectiveness of different standards generally applicable regardless of the state or project;
- Standards would help refocus RRR expenditures on more safety cost-effective geometric improvements; and
- Standards would simplify parts of the design process and FHWA approval procedures, freeing design resources for the analysis of site improvements that cannot be covered by numerical standards.

Lane and shoulder widths on two-lane rural highways meet these conditions, and minimum values are recommended. Cross-section features are particularly important because they can affect highway safety and cost over the length of a highway. Two-lane rural highways account for about three-fourths of all nonfreeway, federal-aid highway mileage and serve roughly one-fourth of all vehicle miles traveled throughout the United States (1).

Where these conditions are not met, for other key features or categories of highways, other design practices are recommended that will help achieve the same safety objectives as minimum standards. In some cases, the recommendations specify threshold conditions that warrant detailed evaluation of particular improvements (Recommendations 5, 7, and 8). In others, they specify improvements that should routinely be made or evaluated on RRR projects (Recommendations 6, 11, and 12), or design policies that should be developed on a state-by-state basis (Recommendations 9 and 10).

Minimum Lane and Shoulder Widths

Recommendation 5: The following minimum lane and shoulder width values are recommended for two-lane rural highways:

Design Year Volume[a] (ADT)	Running Speed[b] (mph)	10 Percent or More Trucks[c]		Less Than 10 Percent Trucks	
		Lane Width	Combined Lane and Shoulder Width[d]	Lane Width	Combined Lane and Shoulder Width[d]
1–750	Under 50	10	12	9	11
	50 and over	10	12	10	12
751–2,000	Under 50	11	13	10	12
	50 and over	12	15	11	14
Over 2,000	All	12	18	11	17

[a]See discussion of design traffic volume later in this chapter (Recommendation 14).
[b]Highway segments should be classified as "under 50" only if most vehicles have an average speed of less than 50 mph over the length of the segment.
[c]For this comparison, trucks are defined as heavy vehicles with six or more tires.
[d]One foot less for highways on mountainous terrain.

The FHWA and state highway agencies can use these recommended minimum lane and shoulder width values to set minimum RRR design standards.

These recommended values are similar to the minimum lane and shoulder width values proposed by the FHWA in 1978 but include several modifications to improve safety cost-effectiveness:

• The average daily traffic (ADT) ranges are adjusted so that a larger number of roads with high ADT and fewer roads with low ADT would be improved. Lane and shoulder width improvements are more cost-effective on high-volume roads than on low-volume roads.
• Minimum roadway widths on roads with high ADT are increased by 2 ft in each direction. The cost per accident eliminated of adding shoulder width on high-volume roads is particularly low.
• Minimum roadway widths on mountainous terrain are reduced by 1 ft in each direction because shoulder width improvements are, on the average, less cost-effective on mountainous terrain.

In terms of cost per accident eliminated, the recommended values are more safety cost-effective than other standards proposed for nationwide use—about $32,000 for each accident eliminated on average compared with $43,000 for the RRR standards proposed by the FHWA in 1978; $41,000 for new construction standards; and $56,000 for the 1977 AASHTO RRR design guide (Chapter 5).

For all federal-aid, two-lane rural highways combined, the recommended minimum values imply approximately the same overall investment as the FHWA standards proposed in 1978—a total of roughly $13 billion if all of the lane and shoulder improvements were made at current cost levels. Application of these values, however, would eliminate about 10,000 (40 percent) additional accidents annually.

The recommended minimum values are stringent enough to ensure that few cost-effective lane and shoulder improvements that enhance safety are excluded from consideration during design, but are not so stringent as to imply an investment level in geometric improvements substantially beyond current practice.

For high-volume roads, the recommended minimum lane and shoulder widths are generally more stringent than special RRR standards, but less stringent than new construction standards. For low-volume roads, the recommended standards are less stringent than the standards now used in most states.

The recommended minimum lane and shoulder widths explicitly take into account vehicle speed and amount of truck traffic, which influence the benefits of wider lanes and shoulders although no reliable quantitative estimates of their incremental effects are available. Distinctions based on vehicle speeds and truck percentages are common in the RRR standards currently in use in many states. These distinctions recognize that on roads with higher vehicle speeds and a larger number of trucks, wider lanes and shoulders have a greater safety payoff.

Less is known about the safety cost-effectiveness of widening urban and multilane rural highways, and minimum values have not been proposed that highway agencies can adopt as standards. The minimum widths recommended for rural two-lane highways can be used as a guide to safety cost-effective improvements for multilane rural and urban highways. However, routinely upgrading lane and shoulder widths in urban areas to the minimum widths recommended for rural two-lane highways is likely to produce some widening projects that are not safety cost-effective, particularly when physical constraints or high right-of-way costs are involved. In such situations, designers must determine the scope of widening improvements on a case-by-case basis.

Horizontal Curvature and Superelevation

Recommendation 6: Highway agencies should increase the superelevation of horizontal curves when the design speed of an existing curve is below the running speeds[1] of approaching vehicles and the existing superelevation is below the allowable maximum specified by AASHTO new construction policies. Highway agencies should evaluate reconstruction of horizontal curves when the design speed of the existing curve is more than 15 mph below the running speeds of approaching vehicles (assuming improved superelevation cannot reduce this difference below 15 mph) and the average daily traffic volume is greater than 750 vehicles per day.

Neither minimum RRR standards proposed for nationwide use in the past (other than new construction standards) nor special RRR standards currently in use set firm requirements for reconstruction of horizontal curves. Furthermore, where states apply new construction standards to federal-aid RRR projects, design exceptions for substandard curves are common and curve reconstruction is infrequent. However, a number of state highway agencies have recognized the need to evaluate alignment improvements during RRR design and have incorporated requirements similar to the preceding recommendation in their special RRR standards.

Current RRR standards and practices generally emphasize lane and shoulder width improvements and do not pay enough attention to alignment

---

[1] The 85th percentile speed (the speed below which 85 percent of the vehicles travel) of approaching vehicles (before slowing for the curve) should be used for this comparison.

improvements. Review of current state RRR practices revealed that lane and shoulder widening is relatively routine but alignment improvements are uncommon. Moreover, once applicable RRR standards have triggered a lane or shoulder improvement, designers often go beyond the RRR minimum standard and specify enough widening to meet new construction standards (Chapter 2).

Shifting RRR investments toward more horizontal curve improvements is warranted. At traffic volumes greater than 750 vehicles per day, reconstruction of horizontal curves can be more cost-effective than lane and shoulder widening and can reduce vehicle operating costs and travel time. Curve reconstruction is rarely cost-effective at traffic volumes lower than 750 vehicles per day even if construction costs are low and potential user benefits are fully considered (Chapter 5).

Because of the variation in costs (and safety cost-effectiveness), however, minimum geometric standards are inappropriate. The cost of reconstructing similar curves varies widely from site to site because of differences in terrain and right-of-way requirements. Thus requiring highway agencies to evaluate curve reconstruction—estimating the added costs, safety benefits, and other user benefits that would result—is the best means available to emphasize the need for selected curve improvements.

Good design minimizes inconsistencies in highway geometry that require motorists to make abrupt or frequent changes in speed. To identify such inconsistencies on existing horizontal curves, designers must compare the design speed of the curve as presently constructed with some measure of the running speeds of traffic on the highway outside the influence of the curve.

This study recommends use of the 85th percentile running speed, measured where approaching traffic has not yet reduced speed. Designers should also consider whether successive curves should be analyzed singly or as a group.

Recommendation 7: At horizontal curves where reconstruction is unwarranted, highway agencies should evaluate less costly safety measures.

Such measures include widening lanes, widening and paving shoulders, flattening steep sideslopes, removing or relocating roadside obstacles, and installing traffic control devices, raised pavement markings, and reflective guideposts. At present highway agencies apply these measures inconsistently.

In many cases, safety can be improved at horizontal curves without costly reconstruction. Depending on site conditions, improvements to curves, short of reconstruction, can be an inexpensive and effective means of reducing the severity and frequency of accidents.

Vertical Curvature and Stopping Sight Distance

Recommendation 8: Highway agencies should evaluate the reconstruction of hill crests when (a) the hill crest hides from view major hazards such as intersections, sharp horizontal curves, or narrow bridges; (b) the average daily traffic is greater than 1,500 vehicles per day; and (c) the design speed of the hill crest (based on the minimum stopping sight distance provided) is more than 20 mph below the running speeds[2] of vehicles on the crest.

Vertical curves are seldom reconstructed to increase stopping sight distance on hill crests. Neither minimum standards proposed earlier for nationwide use by FHWA nor special RRR standards adopted for use in particular states set firm requirements for stopping sight distance. This recommendation is similar to Recommendation 7 (horizontal curve improvements) in that it requires designers to evaluate an improvement when there is a reasonable chance that it will be safety cost-effective.

The study revealed that reconstruction of vertical curves may be cost-effective at average daily traffic volumes greater than 1,500 vehicles per day depending on site conditions. Generally, to be safety cost-effective, vertical curve reconstruction must correct a substantial sight distance restriction that affects drivers' ability to anticipate a hazardous situation—turning vehicles, sharp curves, or other conditions that demand specific driver responses. Unlike horizontal curve reconstruction, hill crest improvements do little to reduce user costs; therefore, reconstruction must be justified primarily on the basis of safety.

Whether or not an evaluation of reconstruction is required, designers should routinely examine the following: the nature of potential hazards hidden by a hill crest, the location of the hazard in relation to the portion of the highway where sight distance falls below the AASHTO new construction standard, and other options such as relocating or correcting the hazard or providing warning signs.

Bridge Width

Recommendation 9: Highway agencies should evaluate bridge replacement or widening if the bridge is less than 100 ft long and the usable width of the bridge is less than the following values:

---

[2] The 85th percentile speed of vehicles passing over the crest should be used for this comparison.

Design Year Volume (ADT)	Usable Bridge Width (ft)[a]
0–750	Width of approach lanes
751–2,000	Width of approach lanes plus 2 ft
2,001–4,000	Width of approach lanes plus 4 ft
Over 4,000	Width of approach lanes plus 6 ft

[a] If lane widening is planned as part of the RRR project, the usable bridge width should be compared with the planned width of the approaches after they are widened.

In most states, highway agencies generally do not widen bridges as part of a RRR project. Special state RRR standards often do not address bridge width requirements, but RRR standards proposed by FHWA in 1978 and the 1977 AASHTO *Geometric Design Guide for Resurfacing, Restoration and Rehabilitation of Highways and Streets (2)* both address minimum bridge widths.

The safety cost-effectiveness of bridge width improvements depends on the usable width of the bridge, the width of approach lanes, traffic volumes, and the length of the bridge (costs for replacement or widening will vary in proportion to length). Designers following this recommendation will analyze bridge replacement or widening in most situations in which bridge width improvements might be justified on the basis of safety cost-effectiveness (Chapter 5).

Recommendation 9 falls between the proposed 1978 FHWA standards for bridges on RRR projects and AASHTO new construction policy for bridges to remain in place on highway reconstruction projects. Under the proposed 1978 FHWA standards, bridges as wide as or wider than the approaches could remain in place regardless of average daily traffic. Otherwise, an evaluation of widening or replacing the existing bridge was required. Under the AASHTO policy, in order for bridges to remain in place when arterial highways are being reconstructed, they should be at least 4 ft wider than the approaches and should be considered for ultimate widening or replacement if their usable width is not at least 6 ft greater than the approaches. Like the proposed 1978 FHWA standards, Recommendation 9 requires highway agencies to evaluate bridge widening when the existing width is less than specified values. At low traffic volumes the recommended values are similar to those proposed by the 1978 FHWA standards, and at high traffic volumes they are similar to those specified by the AASHTO policy for bridges to remain in place on arterial highways.

When evaluating bridge replacement or widening, highway agencies should estimate the following:

- Cost of replacing the existing bridge with a wider bridge designed to AASHTO standards for new bridges,
- Cost of widening the existing bridge (if widening is practical), and
- Number of accidents that would be eliminated by replacement or widening (if widening is practical) (see Chapter 3).

Whether or not evaluation of bridge widening is warranted, designers should routinely consider installing transition guardrails at bridge approaches, rehabilitated or new bridge rails, and warning signs.

Sideslopes and Clear Zones

Recommendation 10: State highway agencies should develop consistent procedures for evaluating and improving roadside features with the following objectives:

- *Flatten sideslopes of 3:1 or steeper at locations where run-off-road accidents are likely to occur (e.g., on the outside of sharp horizontal curves);*
- *Retain current slope widths (without steepening sideslopes) when widening lanes and shoulders unless warranted by special circumstances; and*
- *Remove, relocate, or shield isolated roadside obstacles.*

Neither the RRR standards proposed by AASHTO nor the standards proposed by the FHWA address sideslopes or set clear zone width requirements. However, new construction standards adopted by state highway agencies commonly address both of these roadside characteristics, and a number of states have incorporated numerical sideslope or clear zone width requirements into their special RRR standards approved by the FHWA.

Accident data firmly establish that roadside characteristics are important in determining the overall level of safety provided by a highway. Accident rates are lower and accidents are less severe on highways with few obstacles near the roadway (Chapter 3). Studies of vehicle encroachments onto roadsides and estimates of roadside accident severities, coupled with typical unit costs, indicate that removing isolated trees and relocating utility poles can be more safety cost-effective than widening lanes or flattening horizontal curves (Chapter 5).

Despite these findings, the study revealed no basis for nationwide standards addressing either sideslopes or clear zone width. The safety cost-effectiveness of particular roadside improvements appears highly dependent on site-specific conditions, including not only conditions affecting cost but also interactions between different roadside features that influence the safety benefits of a particular improvement.

Instead of proposing nationwide standards, Recommendation 10 requires highway agencies to develop and apply their own procedures for identifying and selecting sideslope and clear zone width improvements on RRR projects. In some cases, highway agencies might conclude that numerical standards are the best approach for improving safety and promoting efficient design. They could either adopt numerical standards for maximum sideslopes and minimum clear zone width, as some state highway agencies have done already, or they could adopt more flexible design policies that place additional responsibility on the designer. Whichever approach is taken it should reflect the following:

- Wherever possible, sideslopes should not be steepened when widening lanes and shoulders. When the initial slopes are relatively flat, however, the slope can be steepened to 6:1 with little effect, and steepening to 4:1 may be reasonable.
- Permissible sideslopes can be linked to fill heights, as they are by AASHTO new construction standards, recognizing that as fill height increases, sideslope improvements become less cost-effective. Similarly, sideslope requirements could be more stringent as traffic volumes increase because the cost-effectiveness of flatter sideslopes generally increases as traffic volumes increase.
- The cost-effectiveness of removing or relocating a roadside obstacle depends on several factors, including *(a)* the distance of the obstacle from the roadway edge (the shorter the distance, the greater the safety cost-effectiveness); *(b)* the presence of other obstacles nearby (removing an isolated obstacle is more cost-effective than removing an obstacle located between or in front of other obstacles); *(c)* sideslopes on which the obstacle are located (removing an obstacle on a gentle, traversable sideslope is generally more safety cost-effective than removing an obstacle on a steep slope); *(d)* alignment (clear roadsides are more important at curves); *(e)* traffic volume (the greater the traffic volume, the greater the safety cost-effectiveness); and *(f)* speed (the higher the running speeds, the greater the safety cost-effectiveness).
- Clear zone width policies can be tailored to particular types of obstacles commonly encountered in a state to reflect differences in the costs of removal, relocation, or shielding.

Pavement Edge Drop and Shoulder Type

Recommendation 11: To reduce pavement edge-drop hazards on highways with narrow unpaved shoulders, highway agencies should either

- *Selectively pave shoulders at points where out-of-lane vehicle encroachments and pavement edge-drop problems are likely to develop (e.g., at horizontal curves); or*
- *Construct a beveled or tapered pavement edge shape at these points.*

The FHWA currently sets no nationwide standard or specification for type of shoulder construction, for either new construction or RRR, nor does it prescribe requirements for edge shape on resurfacing projects.

Pavement edge drops (i.e., vertical drops or ruts) often develop between the pavement surface and adjacent unpaved shoulders or roadsides. These drops can prevent drivers whose vehicles cross over the lane edge from successfully returning to their original lane without encroaching on an opposing lane or losing control. In addition, pavement edge drops are a common source of tort claims against highway agencies (Chapter 6).

Research sponsored as part of this study indicates that pavement edge drop hazards are greater than previously believed (Chapter 3). However, no basis exists for estimating how often pavement edge drops contribute to accidents or the cost and safety trade-offs involved in preventing or correcting them.

The FHWA requires that edge-drop problems be corrected on completed federal-aid RRR projects at the time of final inspection, but depending on the type of shoulder construction used, resurfacing can increase the likelihood that edge drops will develop later and require repeated maintenance to correct. Shoulders constructed of gravel, turf, or earth are the most susceptible to edge-drop problems; paved shoulders mitigate the problem by moving the edge drop from the lane edge to the outside shoulder edge. On roads with wide unpaved shoulders, paving 2 to 3 ft of the shoulder adjacent to the through traffic lane and striping the edge of the through traffic lane works well at a cost substantially below that of full shoulder paving. In addition, at a given edge drop height, test results show that drivers can recover much more easily when the edge shape is tapered instead of vertical (Chapter 3).

RRR projects can reduce the potential for edge drop-related accidents. Recommendation 11 gives highway agencies considerable latitude in deciding how this can best be accomplished.

Intersections

Recommendation 12: State highway agencies should develop consistent procedures and checklists for evaluating intersection improvements on RRR projects.

Neither minimum RRR standards proposed by FHWA and AASHTO for nationwide use nor most special RRR standards currently in use address intersections. Nevertheless, accidents tend to be concentrated at intersections—more than one-half of all accidents in urban areas and about one-third in rural areas occur at intersections *(3)* (Chapter 3).

Reliable information about the cost and safety trade-offs of individual intersection improvements is generally unavailable because of the large number of physical and operational features affecting intersection safety and because intersection projects typically address multiple intersection safety problems simultaneously. Nevertheless, many intersection improvements can be made at relatively low cost and are safety cost-effective, particularly as traffic volumes increase.

Designers must tailor intersection improvements to site-specific conditions and rely heavily on professional judgment and experience. Useful procedures for selecting safety improvements at intersections include the following:

- Collision diagrams showing vehicle paths, time of occurrence, and weather conditions for individual accidents;
- Condition diagrams showing important physical features that affect traffic movement at the intersection; and
- Field review of the intersection to detect hazards not apparent from the collision and condition diagrams.

Although numerical standards for RRR projects are inappropriate, state highway agencies should develop criteria for identifying intersections that warrant careful evaluation and checklists of improvements to be considered. The criteria might encompass accident frequencies and rates, traffic values, design characteristics, and type of existing traffic control. Improvements could be organized on the basis of three primary design objectives: *(a)* reduction of potential vehicle conflicts (e.g., traffic signals and turning lanes); *(b)* improvement of driver decision-making (e.g., longer lines of sight and lane markings); and *(c)* improvement of the braking capability of vehicles in the intersection (e.g., warning signs to reduce approach speed and increased pavement skid resistance).

Normal Pavement Crown

Recommendation 13: On resurfacing projects, highway agencies should construct pavement overlays with normal pavement crowns that match new construction standards.

Both the earlier AASHTO and FHWA proposals for nationwide RRR standards required that the normal pavement crown—cross slopes from the centerline on straight sections of two-lane roads that allow rainfall to drain to the roadside—be restored to generally match new construction requirements. Resurfacing projects provide highway agencies the opportunity to correct deficient cross slopes at little or no additional cost. Although the safety effects have not been measured, restoring cross slopes to match new construction standards is a good practice that highway agencies should routinely follow when resurfacing.

OTHER DESIGN PROCEDURES AND ASSUMPTIONS

Different highway agencies sometimes design RRR projects differently even when their minimum RRR standards are the same and project conditions are practically identical. Such differences are justified in some cases; for example, one state might have highways with better geometric characteristics and more funds available for RRR work than another. However, differences may occur simply because highway agencies use different procedures and assumptions to apply RRR standards; for example, different assumptions about design traffic volume can change the minimum lane width standard applied.

Procedures are recommended that will encourage a more uniform application of RRR standards and a more consistent approach to safety. Greater uniformity will not prevent highway agencies from tailoring RRR designs to meet the unique conditions of their overall systems or individual projects. State highway agencies have considerable latitude; they can develop their own RRR standards, incorporating the recommendations in the previous section, or seek design exceptions for specific projects.

Design Traffic Volume

Recommendation 14: The design traffic volume for a given highway feature should match the average traffic anticipated over the expected performance period of that feature.

Although projected traffic volumes for some future year are used to select standards for new construction projects, current RRR practice varies among states. Some states use current-year traffic, others use projected future-year traffic. The majority of states, however, use current-year traffic even though the expected performance period of the pavement rehabilitation work is 5 to 15 years and the performance period for geometric improvements may exceed 25 years. The study committee concluded that design decisions for particular highway features should be based on conditions that reflect the anticipated service life of the feature.

Speed

Recommendation 15: When evaluating geometric improvements where vehicle speed is a key factor, highway agencies should estimate running speeds in a manner appropriate for the feature under consideration.

Review of highway agencies' practices revealed that most agencies select a single "design" speed for a RRR project based on highway type, terrain, or the posted speed limits. The FHWA currently requires that the design speed on federal-aid RRR projects equal or exceed the posted or regulatory speed limits.

However, the appropriate speed measure for design varies depending on the feature under consideration. When selecting minimum lane widths, speed should be handled differently than when evaluating a horizontal curve. Furthermore, unlike new construction, RRR project design can use speed parameters based on actual running speeds. Accordingly, the design practices for key features include recommendations on how speed should be taken into account.

When selecting minimum lane and shoulder widths, designers should use a measure of average running speed throughout the project. For horizontal curves, designers should use the 85th percentile speed of vehicles approaching the curve, estimated at a point where drivers have not yet reduced speed. For vertical curves at hill crests where stopping sight distance is limited, designers should use the 85th percentile speed measured on the hill crest.

Pavement resurfacing reduces surface roughness and improves ride quality, which in turn may lead to increases in average speed (Chapter 3). However, these speed increases are usually slight unless the pavement was seriously deteriorated before resurfacing. Therefore, adjusting measured speeds to account for the effect of resurfacing is seldom necessary in RRR project design.

Design Values

Recommendation 16: Highway agencies should estimate the incremental safety cost-effectiveness of improvements that exceed the minimum standard. Designers should consider overall highway geometry, design of adjacent segments, and expected trends in traffic growth and truck use when selecting design values.

Highway designers use minimum RRR standards, such as minimum lane and shoulder widths, to screen existing highway characteristics to determine if lane or shoulder widening, or both, is required. When an improvement must be made, designers must choose the specific design values to be used, which can range from minimum RRR standards to new construction standards. Many highway agencies choose design values based on new construction standards, reasoning that once an improvement has to be made it is sensible to use new construction standards in order to reduce the need for future improvements. However, improvements beyond the RRR minimum standards may not be safety cost-effective and may create inconsistencies between the level of safety provided by the features improved on the RRR project and the features not improved. For example, upgrading cross-section geometry to new construction standards may lead drivers to expect new construction conditions even though commensurate alignment improvements were not made.

Designers are encouraged to make a more deliberate selection of design values and explicitly address issues of incremental safety cost-effectiveness and overall highway consistency in geometric design.

In evaluating incremental cost and safety trade-offs, highway agencies sometimes must make difficult judgments about how much additional cost should be incurred to improve safety. The amount of money one state agency is willing to spend to eliminate an accident may be inappropriate in another state because of differences in existing highway conditions and the financial resources available for systemwide improvements. For rural highways, which typically have greater proportions of more severe accidents, the study committee concluded that when an improvement can eliminate an accident at a cost less than $10,000, it is usually safety cost-effective, but when the cost exceeds $50,000 the improvement is seldom cost-effective. An improvement may or may not be judged safety cost-effective depending on factors such as systemwide safety needs, available financial resources, and the assumed monetary values assigned to highway fatalities, injuries, and property damage.

Design Exceptions

Recommendation 17: When a highway agency requests an exception to a standard, the request should explicitly address the expected safety consequences, along with cost and other impacts.

The review of RRR practices conducted at the outset of this study revealed that the cited justifications for exceptions to design standards were often imprecise and varied from state to state, indicating some confusion over legitimate grounds for design exceptions. To correct this situation, the FHWA has detailed specific requirements for design exceptions consistent with this recommendation.

The design practices recommended in this study should reduce the frequency of design exception requests, but site-specific circumstances will arise for which design exceptions are justified.

PLANNING AND PROGRAMMING RRR PROJECTS

Highway agencies select RRR projects primarily on the basis of pavement repair needs and seldom consider safety needs until preliminary design begins. Given current budget levels and existing highway conditions, pavement repair needs will continue to be the dominant factor in the selection and scheduling of RRR projects. Nevertheless, highway agencies can take safety into account earlier in the overall RRR process in practical ways that will lead to safer highways.

Recommendation 18: Highway agencies should screen the existing characteristics of highways programmed for RRR projects to identify locations where desirable geometric improvements would require additional right-of-way. For such cases, highway agencies should expedite design to determine actual right-of-way requirements and schedule acquisition of the necessary real estate so that it will be available when needed.

After cost, the time required for right-of-way acquisition is the major obstacle to RRR geometric improvements such as reconstruction of horizontal curves, which frequently require additional right-of-way. In a design process geared to completing resurfacing projects from design through construction in one year, additional time is usually not available. Consequently, when right-of-way is needed for a geometric improvement under consideration for a RRR project, highway agencies often must either delay the project or neglect the geometric improvement.

The problem of acquiring right-of-way could be avoided if highway agencies expedited right-of-way acquisition for RRR projects. Although right-of-way acquisition problems are the principal concern, the recommended project screening could be extended to address adverse aesthetic or community impacts that occasionally delay or prevent geometric improvements. Highway agencies could also screen projects for these impacts and work in advance with affected parties to develop an acceptable balance between environmental and safety concerns.

Recommendation 19: Highway agencies should periodically assess the systemwide potential for improving safety through upgraded design.

State highway agencies do relatively little safety planning on a systemwide basis, particularly with respect to geometric design. Safety-oriented planning is usually confined to the federal hazard elimination program for which state highway agencies identify high-accident locations and use federal aid to undertake mainly spot improvements (Chapter 2). Statewide highway needs studies, which many states conduct periodically in one form or another, usually mention safety and may include safety in some manner in developing "sufficiency ratings" of existing highways, but these studies tend to focus on capacity and pavement preservation issues. Thus, most highway agencies have not determined where geometric improvements to existing highways would have the greatest safety payoffs and where such improvements would be the most cost-effective. Although the absence of safety-geometric design relationships and other analysis procedures may have handicapped such efforts in the past, the safety cost-effectiveness analyses conducted in this study illustrate that statewide analysis is practical for some key geometric features such as lane and shoulder widths.

Assessment of the safety cost-effectiveness of making geometric improvements on a statewide basis could increase the positive impact of RRR work on safety in several ways.

- The results would detect any unusual opportunities for safety cost-effective geometric improvements (e.g., highways with narrow lanes and shoulders and high traffic volumes) that might warrant earlier RRR project programming.
- The results could be used to help tailor design practices and standards to the circumstances of a particular state.
- Along with statewide analyses of capacity and preservation needs, the results could be a basis for establishing future state highway programs and funding requirements.
- The assessment could be linked to reviews of other state safety programs (hazard elimination, special state safety programs, railroad grade-crossings, seat belt laws, etc.) to gauge overall progress toward improving highway safety.

Systemwide assessments could also serve as the basis for agreements between state highway agencies and FHWA division offices, such as those discussed in Chapter 2 for California and Ohio, where state spending on special safety improvement programs is linked to requirements for upgraded safety in RRR projects. In California, the state highway agency has earmarked funding for a statewide priority bridge rail upgrading program in lieu of any requirement in the state's federal-aid RRR standards that bridge rail automatically be upgraded in the course of RRR work. Ohio has a similar arrangement with its FHWA division office for guardrail improvements.

SAFETY RESEARCH AND TRAINING

The preceding study recommendations are aimed principally at establishing a more safety-conscious design process. They call for better safety engineering in the design of RRR projects. Better safety engineering requires knowledge about the safety effects of design opportunities and choices, designers with the training and methods to apply this knowledge, and, finally, enough resources devoted to design to permit a thorough design process on each project.

Despite more than one-half century of modern road building, knowledge of the safety consequences of highway design decisions is limited. Furthermore, designers often lack the capability or time to apply the existing knowledge. A clear need exists to expand the knowledge about the relationships between safety and highway design so that designers will be better able to identify safety problems and select cost-effective solutions. Equally important, the highway community must ensure that such knowledge is translated into appropriate methods, manuals, and design aids and that designers receive the safety engineering education and training necessary to apply such design tools.

Taking these steps will mean that additional resources will be needed for safety research, training, and design. The study committee believes the payoff in long-term highway safety gains will be worth the added cost.

The recommendations that follow offer the first step toward meeting long-term research and training needs (Recommendation 20) and also suggest steps that can be taken to improve research and training (Recommendations 21-23).

Recommendation 20: Congress should direct the Secretary of Transportation to establish a special task force to assess highway safety engineering needs and to establish research, education, and funding priorities.

The Interstate highway system is nearly complete, and the United States has shifted emphasis from building new highways to repairing and rebuilding existing highways. As the country proceeds with this enormous task and invests billions of dollars, improvements can be made to the highway system that will reduce traffic fatalities and injuries for decades to come.

The recommended task force should outline research and education agendas to increase safety engineering knowledge. In addition, the task force should consider the question of how well design resources, in terms of staff and funding, are matched to the task of incorporating more extensive safety engineering in highway design and how to promote the use of rigorous statistical controls in safety research.

Recommendation 21: The Federal Highway Administration should develop, distribute, and periodically update a compendium that reports the most probable safety effects of improvements to key highway design features.

Critical reviews of existing safety research are generally unavailable, and organizations such as FHWA and AASHTO, which have a major interest in highway design, have not reported most probable relationships that could be used by designers and others interested in the trade-offs between safety and cost in highway design.

The primary objective of the recommended compendium is to provide designers with the best available safety data and simple application methodologies. Its contents might include the following:

- Background information on the use of accident data and accident models to estimate the safety effects of highway design improvements;
- Easy-to-apply procedures for estimating the safety effects of improvements to specific design features, including a description of the information needed to apply the procedures; and
- How to use the estimates of safety effects, together with information on costs and other impacts, to assess the cost-effectiveness of design improvements.

The recommended compendium would build on earlier research reviews and the findings of this study and would provide designers a common starting point for their work, which could then be adjusted to take into account local accident histories or other highway circumstances. In addition, a safety compendium could help redirect current research efforts and focus attention on the value of research in at least three ways:

- Designers will be able to apply research results sooner because *(a)* results would be keyed to most probable relationships contained in the compendium and *(b)* the compendium would be periodically updated to incorporate new research findings.
- Major sponsors of safety research (FHWA, AASHTO, and individual state highway agencies) will identify the principal gaps in current knowledge and determine where research might be able to fill these gaps.
- Researchers will have a frame of reference for conducting their work; study objectives could be aimed, for example, at validating a safety relationship, testing an assumption in a recommended procedure, or filling a key gap in current knowledge.

Recommendation 22: The Federal Highway Administration and the National Cooperative Highway Research Program (NCHRP) should increase research on the relationships between safety and highway design.

Except for a modest FHWA research program and occasional NCHRP-sponsored research studies, few opportunities exist for coordinated, purposeful safety research aimed at highway design. To a large extent, the highway community has relied on uncoordinated research without rigorous statistical controls to expand knowledge about the safety effects of road design. Certainly some state highway agencies have evaluated the safety impacts of various types of highway improvements and reached conclusions that have been usefully applied in subsequent design work. However, as a rule, such efforts lack the statistical controls necessary to develop relationships that can be reliably transferred to other locations or generalized for nationwide application.

Significant progress will come only through research programs that carefully define objectives, select an appropriate experimental setting, collect the necessary data, and apply appropriate analysis techniques. Safety research is especially complex because of the number of factors other than highway design responsible for accidents. Research projects do not always fully achieve their objectives, but when they do the benefits can be substantial.

Although a new study of research priorities is recommended, existing FHWA and NCHRP programs can begin to increase knowledge of the relationships between safety and highway design. Especially important topics that merit further research include

- Safety effect of physical and operational features of intersections,
- Safety effects of lane and shoulder conditions on urban highways and streets,
- Safety effects of different sideslope and other roadside conditions,
- Safety effects of low-cost safety treatments such as warning devices at hazardous locations or shoulder widening at horizontal curves, and
- Combined safety effects of changes in horizontal and vertical alignment.

Recommendation 23: The Federal Highway Administration, the American Association of State Highway and Transportation Officials, state and local highway agencies, and other organizations of public works professionals should support continuing training activities to keep design engineers abreast of safety-conscious design.

Regardless of how RRR projects are selected, which standards are applied, or how the design process is organized, individual design engineers ultimately play the major role in determining the degree to which RRR projects will enhance safety. Although engineers can rely on standards and guidelines to make many design decisions, some decisions must be based on site-specific circumstances and judgment. The quality of these decisions depends on how well the engineer is prepared for designing RRR projects.

The special task force recommended (Recommendation 20) should address the long-term needs for greater safety skills and awareness among designers through university curricula and in-service training. In the meantime, FHWA, AASHTO, and individual highway agencies can use a combination of formal and informal training techniques to increase the skills of design engineers. Formal training includes short courses and seminars similar to those conducted through FHWA's National Highway Institute. Informal training could include reviews of completed RRR projects or conferences that provide forums for designers to share experiences. Both types of activities can be effective, and can offer highway agencies a range of options to fit time and budget constraints.

REFERENCES

1. *Highway Statistics 1985*. FHWA, U.S. Department of Transportation, 1986.
2. *Geometric Design Guide for Resurfacing, Restoration and Rehabilitation of Highways and Streets*. American Association of State Highway and Transportation Officials, Washington, D.C., 1977.
3. *Accident Facts—1985 Edition*. National Safety Council, Washington, D.C.

Appendix A
Summary Comparison of Nonfreeway Geometric Design Standards and Guidelines

Table A-1 contains a description of three sets of geometric design standards and guidelines for nonfreeway highways. The AASHTO RRR guidelines are from the American Association of State Highway and Transportation Officials' *Geometric Design Guide for Resurfacing, Restoration, and Rehabilitation (RRR) of Highways and Streets (1)*. This guide, commonly referred to as the "purple book," contains minimum design values for lane and shoulder widths, cross slopes, superelevation, and bridge widths, as well as advisory information on grades, curvature, sight distance, and clear zones. Overall, the AASHTO RRR guidelines are considerably less stringent than new construction standards.

In August 1978, after opposition to the AASHTO RRR guidelines surfaced, the Federal Highway Administration (FHWA) proposed RRR standards *(2)*. These proposed standards are generally more stringent than the AASHTO RRR guidelines, but still less stringent than new construction standards.

In June 1982, the FHWA issued regulations permitting states to develop their own RRR standards, subject to FHWA approval. However, some states continue to use new construction standards for RRR projects, with design exceptions on a case-by-case basis. AASHTO's *Policy on Geometric Design of Highways and Streets, 1984 (3)* has been approved by the FHWA for the design of new and major reconstruction federal-aid projects. The policy also is applicable to RRR work in states that use new construction standards for RRR projects.

TABLE A-1 Summary Comparison of Nonfreeway Geometric Design Standards and Guidelines

	AASHTO RRR Guidelines	1978 FHWA RRR Proposed Standards	AASHTO Policy for New Construction
Traffic data (current)	ADT, DHV, percent trucks, and turning movements at signalized intersections must be collected and analyzed.	ADT, DHV, percent trucks, accident locations, and descriptions, including collision diagrams, should be collected and analyzed. At signalized intersections, turning movements and pedestrian volumes should be collected and analyzed.	ADT, DHV, directional distribution, traffic composition (percent trucks)
Future traffic projections	5 to 10-year traffic forecasts for major rehabilitation.	Not specified.	Design year 20 years ahead "widely used"; design year 5 to 10 years may be appropriate for reconstruction and rehabilitation projects.
Design speed	No minimum, but design should accommodate current running speed.	Average free-flow running speed plus 10 percent.	*Rural* Arterials: 50–70 mph depending on terrain. Collectors: 20–60 mph depending on terrain, ADT, DHV. Local: 20–50 mph depending on terrain, ADT, DHV. *Urban* Arterials: "generally" 40–60 mph; "occasionally" 30 mph under restricted conditions. Collectors: 30–60 mph depending on terrain, traffic, intersection spacing. Local: 20–30 mph depending on terrain, traffic, development.

TABLE A-1 continued

	AASHTO RRR Guidelines	1978 FHWA RRR Proposed Standards	AASHTO Policy for New Construction
Superelevations	Use new construction rates unless constraints do not permit. Use 10 degrees BBI reading to determine maximum safe speed.	Rates for new construction apply. Design speed may be reduced if necessary, but special signing required. Such reductions considered infrequent for rural areas; 10 degrees BBI reading used to establish safe speed.	*Rural:* function of DS, terrain, climate; maximum 0.10; 0.08 where snow and ice are factors. *Urban:* 0.04 to 0.06 maximum on higher speed streets with few restrictions; generally no superelevation on low-speed curbed streets.
Rural two-lane			
Minimum lane and shoulder widths			
ADT = 350 DS = 40 mph	10 ft (2 ft)	10 ft (2 ft) 9 ft (2 ft) if "minor road"	Local: 10 ft (2 ft) Collector: 10 ft (2 ft, 4 ft if barrier) Arterial: NA (arterial recommended DS ≥ 50 mph)
ADT = 500 DS = 40 truck > 10 percent	10 ft (2 ft)	11 ft (2 ft)	Local: 11 ft (4 ft) Collector: 11 ft (4 ft, 6 ft) Arterial: NA
ADT = 500 DS = 40 truck < 10 percent	10 ft (2 ft)	10 ft (2 ft)	Local: 11 ft (4 ft) Collector: 11 ft (4 ft, 6 ft) Arterial: NA
ADT > 400 DHV > 400 truck > 10 percent	10 ft (2 ft)	12 ft (4 ft)	Local: 12 ft (8 ft) Collector: 12 ft (8 ft) Arterial: 12 ft (10 ft) (for all design speeds)

TABLE A-1 *continued*

	AASHTO RRR Guidelines	1978 FHWA RRR Proposed Standards	AASHTO Policy for New Construction
Rural multilane Minimum lane and shoulder widths			
DS ≤ 50 and trucks < 10 percent	10 ft (2 ft)	10 ft (2 ft)	12 ft (10 ft-shoulder "preferable," 8-ft minimum)
DS > 50 or trucks ≥ 10 percent	10 ft (2 ft)	11 ft (4 ft)	12 ft (10 ft-shoulder "preferable," 8-ft minimum)
Urban arterials Minimum lane widths			
Through lanes	10 ft	10 ft	10 ft, highly restricted conditions, low truck traffic 11 ft "adequate" 12 ft "desirable," "generally used" on higher speed (≥ 40 mph), free-flowing principal arterials.
Parking lanes	7 ft	8 ft	10–12 ft (8 ft acceptable if never used as a traffic lane).
Turning lanes			
DS ≤ 40 mph	9 ft	10 ft	10-ft left-turn lanes, 11-ft continuous two-way, left-turn lane.
DS > 40 mph	10 ft	10 ft	10-ft left-turn lanes, 11-ft continuous two-way, left-turn lane.

TABLE A-1 continued

	AASHTO RRR Guidelines	1978 FHWA RRR Proposed Standards	AASHTO Policy for New Construction
Horizontal curvature, maximum grade, and minimum stopping sight distance	Improvement should be considered at high-accident locations. Considerable engineering judgment must be exercised. Use signing when sight distance less than AASHTO standard for new construction.	When curve DS is no more than 15 mph less than roadway DS, signing required. When curve DS exceeds roadway DS by more than 15 mph, corrective work should be undertaken unless "impractical."	Minimum radius a function of design speed, maximum superelevation, and side friction. Maximum grade a function of terrain, design speed, percent trucks, traffic volume, functional class: Rural (%) Urban (%) Local 5–16 4–15 Collector 4–12 5–14 Arterial 3–7 5–11 Minimum stopping sight distance a function of design speed and grade. On level wet pavement: DS, mph Assumed speed, mph Minimum stopping distance, ft 30 28–30 200–200 50 44–50 400–475 70 58–70 625–850
Bridges Minimum width (existing bridges)	No absolute minimum specified. One-way operation permitted. Minimum guideline of 18 ft provided for low-volume (ADT ≤ 250) minor roads with few trucks.	18 ft for bridge requiring minor rehabilitation. 20 ft for low-volume (ADT ≤ 250) bridges requiring major rehabilitation.	*Collectors* DHV Width < 200 22 ft 200–400 24 ft > 400 28 ft *Rural arterials* Travel lanes plus 2 ft each side *Urban arterials* Curb-to-curb width

TABLE A-1 continued

	AASHTO RRR Guidelines	1978 FHWA RRR Proposed Standards	AASHTO Policy for New Construction
Clear zone			
Rural	From edge of pavement, 30 ft desirable, but there must be many exceptions. Emphasis on removing fixed objects identified as hazardous by accident analyses.	Review accident data to define dangerous obstructions. Considerable judgment must be used because of existing topographic and right-of-way limitations.	Arterials and high-speed ($\geq$ 50 mph DS) collectors: follow 1977 AASHTO *Guide for Selecting, Locating, and Designing Traffic Barriers (4)*. All fixed objects should be outside clear zone.[a] Low-speed collectors, local: minimum 10 ft; exception may be made where guardrail provided. Consider safety, environmental, aesthetic concerns in deciding whether to remove all trees, poles, and so forth.
Urban	Minimum setback should be behind the pared shoulder or curbing.	Minimum setback should be behind the pared shoulder or 2 ft behind the curb.	With curbs: 1.5 ft beyond face of curb; without curbs: same as rural. Essential to remove only "very vulnerable" fixed objects on urban collectors.
Safety appurtenances	Provide traffic barriers for clear zone hazards that cannot be eliminated.	Desirable to upgrade to current criteria as part of RRR project.	Follow 1977 AASHTO *Guide for Selecting, Locating, and Designing Traffic Barriers (4)*

NOTE: ADT – average daily traffic; DHV – design hour volume; DS – design speed; BBI – ball bank indicator; and NA – not applicable.
[a]Clear zone as a function of DS, slope, and curvature.

REFERENCES

1. *Geometric Design Guides for Resurfacing, Restoration, and Rehabilitation (RRR) of Highways and Streets.* American Association of State Highway and Transportation Officials, Washington, D.C., 1977.
2. "Design Standards for Highways, Notice of Proposed Rulemaking." *Federal Register,* Vol. 43, No. 164, Aug. 23, 1978.
3. *A Policy on Geometric Design of Highways and Streets.* American Association of State Highway and Transportation Officials, Washington, D.C., 1984.
4. *Guide for Selecting, Locating and Designing Traffic Barriers.* American Association, of State Highway and Transportation Officials, Washington, D.C., 1977.

Appendix B
Case Study State and Local RRR Programs

The tables in this appendix contain a description of resurfacing, restoration, and rehabilitation (RRR) finance and expenditures; programming methods; and design standards and practices in the state highway agencies and local governments chosen as case studies. The case studies are discussed in detail in Chapter 2.

Most of the information presented in the tables was assembled in 1984 and was the most recent available at the time. However, Tables B-9 and B-10 give the special RRR design standards in effect in October 1986 in the case study states.

Follow-up interviews were conducted with highway agency officials in each of the 15 case study states in the fall of 1986. Officials were asked to list the changes that had occurred in their RRR programs since the original interviews. The follow-up interviews indicated no major changes had occurred in the states' RRR activities in the 2-year period and that trends observed in 1984 have continued. Several states reported that emphasis on safety in RRR projects has continued to grow. Some changes have been made in RRR standards, and although most have been simply fine-tuned, a few states (Ohio and Texas) have made substantial revisions, and one state (Arizona) is in the process of developing special RRR standards for the first time. A number of state officials reported that they have received increased guidance from the FHWA on RRR design matters and that the FHWA has given more attention to reviewing exception requests during the past 2 years. In finance as in design, the states experienced few major changes, although some reported an increase in the amount of money available for fully state-funded RRR projects over the period.

TABLE B-1 Characteristics of the Case Study States

	Federal-Aid System				Program Administration	
State	1984 Mileage (thousands)	Percent Urban	Percent State Administered	Fiscal Year 1984 Federal-Aid Apportionment (millions)	Special RRR Standards Approved	Certification Acceptance for Non-Interstate Primary Design
Arizona	9	22	62	239	No	No
California	41	46	37	907	Yes	No
Florida	20	40	55	419	Yes	No
Illinois	31	28	49	544	Yes	Yes
Michigan	31	21	30	336	Yes	No
Mississippi	20	11	51	152	Yes	No
Missouri	29	11	94	277	No	Yes
New Hampshire	3	30	88	55	No	No
New Jersey	9	69	26	288	Yes	Yes
New York	25	45	60	617	No	Yes
Ohio	27	36	67	396	No	No
South Dakota	18	3	50	78	Yes	No
Texas	60	16	93	796	Yes	No
Virginia	20	18	93	304	No	Yes
Washington	17	27	42	283	Yes	Yes
Total	360			5,691		
U.S. total or average	838	22	63	11,740	26, yes	15, yes

SOURCES: *Highway Statistics 1984*, FHWA, Tables HM-14, HM-15, and FE-221; interviews with highway agency officials in case study states; FHWA records of RRR standards and certification acceptances.

TABLE B-2 Percent of Federal-Aid Highway Miles Under State Administration, by Federal-Aid System, Case Study States

	Federal-Aid Highway System			Total Federal-Aid System
	Interstate and Primary	Secondary	Urban	
Arizona	100	38	6	62
California	98	8	5	37
Florida	94	9	26	55
Illinois	99	14	26	49
Michigan	100	12	3	30
Mississippi	100	30	14	51
Missouri	100	100	29	94
New Hampshire	98	100	38	88
New Jersey	97	5	10	26
New York	95	66	16	60
Ohio	100	73	20	67
South Dakota	99	12	4	50
Texas	100	99	41	93
Virginia	100	99	53	93
Washington	100	15	3	42
United States	98	49	23	63

SOURCE: *Highway Statistics 1984,* FHWA, Table HM-14.

TABLE B-3 Distribution of Non-Interstate Federal-Aid Project Expenditures by Project Category in Case Study States

	Percent of Total Non-Interstate Federal-Aid Expenditures					
State	Reconstruction and New Construction	Resurfacing and Minor Widening	Bridge Work	Safety Improvements	Intersection Improvements	Other	Comments[a]
Arizona	58.0	38.0	–	4.0	–	–	FY 1984 programmed; federal-aid primary and federal-aid secondary only.
California	37.7	32.2	11.8	7.0	7.1	4.2	FY 1985 programmed; excludes local federal-aid secondary and federal-aid urban; based on projects more than $250,000.
Florida	71.8	13.2	14.9	–	0.1	–	FY 1985 programmed; safety and "other" in construction or resurfacing categories; includes local federal-aid secondary and federal-aid urban.
Illinois	46.2	29.0	11.6	8.0	–	5.2	FY 1984 programmed; includes local federal-aid secondary and federal-aid secondary.
Michigan	40.0	33.9	16.2	2.5	3.2	4.2	FY 1984 programmed; does not include local federal-aid secondary.
Mississippi	73.2	11.3	14.9	–	–	0.6	FY 1984 programmed; federal-aid primary and federal-aid secondary only; includes local federal-aid secondary.
Missouri	13.2	13.0	66.9	6.2	0.1	0.6	FY 1984 programmed; derived from data on state-defined systems (state and federal dollars may be mixed).
New Hampshire	28.7	8.9	51.1	2.5	6.6	2.2	FY 1984 programmed; includes local federal-aid urban and federal-aid secondary.

TABLE B-3 *continued*

State	Percent of Total Non-Interstate Federal-Aid Expenditures					
	Reconstruction and New Construction	Resurfacing and Minor Widening	Bridge Work	Safety Improvements	Intersection Improvements	Other	Comments[a]
New Jersey	43.9	12.2	14.4	9.8	18.0	1.7	FY 1984 programmed; includes local federal-aid secondary and federal-aid urban.
New York	22.9	15.5	50.8	2.6	2.3	5.9	FY 1985 programmed; includes local federal-aid secondary and federal-aid urban.
Ohio	20.5	42.4	27.2	8.1	1.3	0.8	FY 1984 contract awards.
South Dakota	32.9	50.5	13.5	0.9	–	2.2	FY 1983 contract lettings; excludes local federal-aid urban.
Texas	67.4	28.8	–	–	0.1	3.7	FY 1983 contract lettings; bridge and safety projects in other categories; includes local projects.
Virginia	59.0	1.0	26.0	–	13.5	0.5	FY 1984 programmed, federal-aid primary and federal-aid urban only; safety projects in other categories.
Washington	29.0	42.6	20.6	6.7	0.7	0.4	FY 1985 programmed; excludes local federal-aid secondary and federal-aid urban.

NOTES: Amounts include state match and construction phases only. Dashes in cell indicate the project category could not be separated from other types in the data supplied by the state. The percentage distributions reflect classification of the entire cost of each project into the one category corresponding to the primary motivation for conducting the project. For example, if a project involving resurfacing, bridge repair, and safety improvements was originally selected for programming because of deteriorated pavement, its entire cost is in the resurfacing and minor widening category, even though the ancillary improvements may have accounted for a substantial portion of that cost.

[a]Expenditures are defined as the full costs of all projects actually commenced in the year, or costs of projects programmed (i.e., planned) for the year. Programmed amounts are estimated expenditures and may not have been incurred in the given fiscal year.

TABLE B-4 Distribution of State DOT Expenditures for Fully State-Funded Projects, Selected Case Study States

State	Reconstruction and New Construction (%)	Resurfacing and Minor Widening (%)	Seal Coats and Thin Overlays (%)	Bridge Work (%)	Safety Improvements (%)	Intersection Improvements (%)	Other (%)	Notes
California	0	7.2	39.2	0	—	0.4	53.2	FY 1985 programmed; includes only projects more than $250,000, except all seal coats included
Illinois	4.7	75.8	—[a]	2.0	7.4	—	10.1	FY 1984 programmed
Michigan	0	60.1	—	0	—	0	39.9	FY 1984 programmed
Mississippi	0.1	72.9	19.3	—	—	0.4	7.3	FY 1984 programmed
New Jersey	29.3	7.7	30.8	0	—	16.4	15.9	FY 1984 programmed
New York	13.6	45.9	—	27.6	—	3.1	9.9	FY 1984 programmed; safety included in Other category
Ohio	0	56.8	—[b]	—	2.9	1.0	39.3	FY 1983 actual contract awards
South Dakota	0	27.4	57.5	2.4	12.8	—	0	FY 1983 actual contract lettings
Texas	51.8	41.0	—[c]	—	—	1.2	6.0	FY 1983 actual contract awards
Washington	65.5	12.4	8.3	7.0	—	2.3	4.6	FY 1985 programmed

NOTES: Expressed as percent of total expenditures for state-funded projects. Amounts in Total Expenditures column include state-funded construction and (unless otherwise noted) costs of seal coats and thin overlays as maintenance activities. Dashes in cells mean that no information is available for that category; expenditures may be included in Other category.

[a] In Illinois, seal coats and thin overlays are a maintenance activity; dollar amount is unavailable and not included in total expenditures.
[b] Seal coats and thin overlays included in Other category.
[c] Seal coats and thin overlays included in Resurfacing and Minor Widening category.

TABLE B-5 RRR Project Programming Procedures in the Case Study States

State	Programming Procedures
Arizona	Central office selects projects using its pavement management system, an automated procedure for determining the least-cost schedule of pavement repairs that will maintain a specified systemwide minimum performance standard. Inputs to the process include annual measurements of pavement deflection, cracking, and roughness.
California	RRR project selection is guided by a pavement management system (PMS), a biennial pavement survey, and priority ranking. Districts recommend major projects (more than $250,000) to central office following PMS priorities; central office assembles a 3-year RRR program based on projects' statewide priority rankings. Ride quality is an important consideration in assigning priorities. Almost all major RRR is federal-aid. Districts receive funding allocations for minor construction RRR and maintenance overlays and select their own projects in these programs, subject to central office review.
Florida	Central office allocates resurfacing funds by formula to each district. Districts receive separate allocations for federal-aid and state-funded resurfacing. Districts select projects guided by PMS ratings, but are allowed to exercise considerable leeway to exercise judgment. Districts choose between federal aid and state funding for each project, considering standards requirements and urgency.
Illinois	District offices annually propose RRR projects for inclusion in the 5-year improvement program based primarily on professional judgment, local knowledge, and a visual pavement condition survey. The central office screens these proposals with respect to pavement condition estimates, traffic levels, and geometric condition, and designates projects that balance statewide needs. District offices may make project substitutions provided overall costs are about the same. In addition, central office undertakes resurfacing projects as part of a mostly state-funded winter damage program. Annually, the central office allocates funds for this program to the districts, where project selections are made.
Michigan	Central office selects RRR projects mainly judgmentally, based on district recommendations and a biennial subjective sufficiency rating of state roads. Formal geographic or program allocations of funds are not used. Nearly all resurfacing is federal-aid. PMS is under development.
Mississippi	Districts determine resurfacing project needs judgmentally, then choose which projects to recommend to the central office as federal-aid RRR construction and which to conduct as state-funded resurfacing using funding allocations for maintenance overlays. Resurfacing-only projects are usually state-funded and federal-aid RRR is usually on roads that need service upgrading in addition to resurfacing. Central office chooses federal-aid RRR projects from among district recommendations on the basis of a statewide priority ranking.

TABLE B-5 *continued*

State	Programming Procedures
Missouri	Central office assigns a mileage allocation to each district. Districts select their "worst miles" of pavement up to their allocations. A central office team inspects and visually rates these worst miles. Districts then program RRR projects within their total construction program funding allocations, following central office rules as to the rating scores that warrant a RRR project. Nearly all construction resurfacing is federal-aid. Maintenance resurfacing (mainly on low-volume roads) is selected by central office maintenance division (for larger projects) or by districts (for small projects) within their maintenance resurfacing funding allocations.
New Hampshire	Federal-aid RRR projects are selected by the central office from a list of candidates compiled at the beginning of the federal-aid RRR program by the state and the FHWA. Most resurfacing currently is state funded and selected judgmentally by the central office maintenance division on the basis of district recommendations.
New Jersey	Central office design division selects and designs more extensive RRR projects, which are usually funded using federal-aid. Projects typically are identified because of needed pavement repairs. The central office maintenance division handles state-funded resurfacing work selected on the basis of systemwide roughness measurements.
New York	Central office undertakes RRR-related work in three programming categories: regular federal-aid capital, state reconditioning and preservation, and transportation improvement materials (maintenance resurfacing). For each, the central office allocates available funds to regions and regional offices then selects projects that match available funds. Project selections are motivated mostly by pavement repair needs that are assessed by professional judgment, local knowledge, and a statewide visual pavement survey.
Ohio	Performs RRR-related work in three programs administered by different central office bureaus: location and design, traffic, and maintenance. In each case, at least some federal aid is used. The central office bureaus establish program guidelines and fund allocations, district offices select individual projects, and finally, the central office bureaus review and schedule all projects. Except for special traffic or safety projects, districts select projects primarily on the basis of pavement repair needs assessed by professional judgment and local knowledge.

TABLE B-5 *continued*

State	Programming Procedures
South Dakota	Central office selects all RRR projects through an annual process that updates the state's 5-year construction program. Project priorities are initially established on the basis of pavement design and condition (from visual surveys, roughness measurements, and deflection measurements), drainage adequacy, and traffic characteristics and adjusted after field reviews. Projects are scheduled in order of priority to match allocations of available funds among different functional classes and between resurfacing and other types of construction work.
Texas	Central office undertakes RRR projects principally in its rehabilitation program category. Normally, the highway commission sets the amount of funding and allocates funds to districts. District offices then select projects that in total match their fund allocations. These project selections are based mostly on professional judgment and knowledge of local conditions and typically address pavement repair needs, as well as geometric deficiencies.
Virginia	Central office performs RRR improvements through its construction and maintenance programs. On the state primary system, the central office selects construction projects, with input from district offices, that in total match funds allocated to each district. On the state secondary system, county governments select projects, in consultation with district offices, that match construction funds allocated to each district. Under the maintenance program, the central office allocates resurfacing funds to each district and district offices then selects projects that match their funding.
Washington	Central office allocates resurfacing funds by formula to districts and also assigns resurfacing priorities based on PMS ratings. Districts assemble resurfacing programs and must address all high priority roads, but can choose between federal-aid-eligible major RRR or state-funded light overlays. About one-half of major RRR projects are also state-funded, but in general federal-aid and state RRR projects are not treated separately during programming.

TABLE B-6 Comparison of Federal and State Funding for Resurfacing, Case Study States

State	Fiscal Year	Federal-Aid Projects' Share of Total State Highway Agency Non-Interstate Resurfacing		Percent of State-Maintained Non-Interstate Federal-Aid Mileage Resurfaced in Fiscal Year, Federal-Aid Projects	Percent of Total State-Maintained Non-Interstate Mileage Resurfaced in Fiscal Year, State and Federal Funding
		Federal-Aid Projects as Percent of Expenditures[a]	Federal-Aid Projects as Percent of Miles Resurfaced		
Arizona	1984	68	31	1.9	6
California[b]	1985	68	20	2.5	12
Florida	1985	26	23	0.9[d]	6
Illinois	1984	19	12	0.7	NA
Michigan	1984	98	99	3.3	3
Mississippi	1984	19	6	1.0	16
Missouri[b,c]	1985	35	7	0.9	12
New Hampshire[b,c]	1985	38	2	0.2	7
New Jersey[b]	1984	78	NA	0.6[d]	NA
New York	1984	43	NA	NA[d]	NA
Ohio	1983	71	62	3.9[d]	NA
South Dakota[b]	1983	82	NA	1.2[d]	NA
Texas[b,c]	1983	40	8	0.5	8
Virginia[b,c]	1985	8	2		
Washington	1984	60	37	3.6	10

NOTE: NA indicates percentage is not available.
[a]State match included in federal-aid expenditures.
[b]More than 50 percent of state resurfacing expenditures are for thin overlays and seal coats costing less than $40,000/mi.
[c]State system includes substantial nonfederal-aid mileage.
[d]Includes mileage and resurfacing of some locally administered federal-aid secondary or urban roads.

TABLE B-7 Funding for Special Safety Improvement Projects, Case Study States

	Ariz.	Calif.	Fla.	Ill.	Mich.	Miss.	Mo.	N.H.	N.J.	N.Y.	Ohio	S.Dak.	Tex.	Va.	Wash.
Federal-aid primary, secondary, or urban funds used for safety projects?	No	Yes	Yes	No	Yes	No	No	No[a]	No	No[a]	Yes	No	No	No	Yes
100 percent state construction funds used for safety projects?	No	Yes	Yes	No	No	No	No	Yes	No	Yes	No	Yes	No	No	Yes
Earmarked maintenance budget item for safety improvements?[b]	No	No	No	Yes	No	No	Yes	No	–[c]	–[c]	Yes	–[c]	–[c]	No	No
Percent of federal hazard elimination funds routinely made available to local governments	25	50	0	0	50	0	0	0	0	0	0	0	0	0	60
Unobligated hazard elimination balance as of June 30, 1987 as percent of annual apportionment	180	63	146	53	0	163	33	129	153	71	104	168	206	41	115

NOTE: Special safety improvement project is primarily motivated by a safety concern and intended to correct a specific hazardous condition.
[a]With rare exceptions.
[b]All case study states conduct safety improvements with maintenance funds; "yes" indicates state has a budgeted special-purpose maintenance program addressing safety.
[c]Data not available.

TABLE B-8 Design Practices for Federal-Aid RRR Projects, Case Study States

	Ariz.	Calif.	Fla.	Ill.	Mich.	Miss.	Mo.	N.H.	N.J.	N.Y.	Ohio	S.Dak.	Tex.	Va.	Wash.
Use special RRR standards?	No	Yes	Yes	Yes	Yes	Yes	No	No	Yes	No	No	Yes	Yes	No	Yes
Most design at district offices?	No	Yes	Yes	Yes	No	No	Yes	No	No	Yes	Yes	No	Yes	Yes	Yes
Certification acceptance for design and construction on primary highways?	No	No	No	Yes	No	No	Yes	No	No	Yes	No	No	No	Yes	Yes
Safety and geometric needs identified and reviewed															
Predesign report submitted to central office or the FHWA	o	+	+	+	o	+	+	x	+	+	x	+	o	o	+
Predesign field review with the FHWA	x	+	x	o	+	+	+	x	x	x	+	o	o	o	x
Review of accident data	+	+	+	+	+	+	+	+	+	+	+	+	x	o	+
Review of state traffic or safety staff	o	o	o	o	+	+	o	o	o	o	o	o	o	x	o
Selected use of formal cost-effectiveness analysis	o	o	o	o	x	o	o	o	o	o	x	x	o	o	o

NOTE: + = always or usually, x = occasionally, and o = rarely or never.

TABLE B-9 Minimum Geometric Requirements for Rural Highways in Case Study States With Special RRR Standards

Condition	California	Florida	Illinois	Michigan
Highway Design Speed (mph)				
40	Not specified	ARS = 40 mph	Where posted speed = 40 mph	35 mph posted
45	Not specified	ARS = 45 mph	Where posted speed = 45 mph	40 mph posted
50	Not specified	ARS = 50 mph	Where posted speed = 50 mph	45 mph posted
Lane width (ft)				
10	_c	ARS = 40 mph ARS = 40, ADT = 1,000 T = 10%	DS = 40 and ADT = 1,000 ADT = 400 (all DS)	Two-lane highways where ADT < 750
11	_c	ARS > 40 mph	DS = 50 and ADT = 3,000 ADT = 1,000 ADT = 3,000 and existing roadway = 30 ft	All highways
Shoulder Width (ft)				
2	_c	Not permitted	DS = 50 and ADT = 250	Not permitted
4	_c	Not permitted	ADT = 3,000	All highways except on two-lane highways where ADT > 10,000

Mississippi	New Jersey[a]	South Dakota[b]	Texas	Washington
ARS = 36 mph	*Resurfacing and Restoration* Posted speed = 40 mph *Rehabilitation* Posted speed = 35 mph	Not specified	Undivided multilane highways in rolling terrain Two-lane highways except in flat terrain with ADT = 1,500	Not specified (should be logical with respect to terrain and type of highway)
ARS = 41 mph	*Resurfacing and Restoration* Posted speed = 45 mph *Rehabilitation* Posted speed = 40 mph	Not specified	Undivided multilane highways in rolling terrain Two-lane highways except in flat terrain with ADT = 1,500	Not specified (should be logical with respect to terrain and type of highway)
ARS = 45 mph	*Resurfacing and Restoration* Posted speed = 50 mph *Rehabilitation* Posted speed = 45 mph	Not specified	All highways	Not specified (should be logical with respect to terrain and type of highway)
Not permitted	Not permitted	Not permitted	Two-lane highways where ADT < 750	Not permitted
All multilane highways Two-lane highways where ADT = 750	All highways for resurfacing and restoration projects	*Two-lane Collectors* Minor reconstruction: ADT < 250 Resurfacing: ADT > 500	All highways	All multilane highways except where ADT > 4,000 and T = 10% All two-lane highways except where ADT ≥ 1,000 and T = 10%
Multilane Highways on left side DS = 50, ADT = 4,000 DS = 50, ADT > 4,000, and T < 10%	On left side of multilane, divided highways	*Principal Arterials* Resurfacing: ADT = 1,500 *Minor Arterials* Minor reconstruction: ADT = 500 All resurfacing projects *Collectors* All resurfacing and minor reconstruction projects	Two-lane highways with ADT < 750	On left side of multilane divided highways On two-lane highways where ADT < 1,000 or DS < 50 mph and ADT < 2,000
All highways	On left side of multilane divided highways	*Principal Arterials* Minor reconstruction: ADT = 1,000 All resurfacing; *Minor Arterials and Collectors* All resurfacing and minor reconstruction	All highways	All two-lane highways Multilane highways where ADT < 4,000

TABLE B-9 *continued*

Condition	California	Florida	Illinois	Michigan
Shoulder Width (ft)				
6	_[c]	All highways	All highways DS > 50 ADT = 5,000	All highways except on two-lane highways where ADT > 10,000
Spot Horizontal Curve Design Speed (mph)				
35	Not specified	ARS = 50 mph[d]	DS = 45 mph	DS = 35-50 mph and not a high-accident location[d]
40	Not specified	ARS = 55 mph[d]	DS = 50 mph	DS = 40-55 mph and not a high-accident location
45	Not specified	ARS = 60 mph[d]	DS = 55 mph	DS = 45-60 mph and not a high-accident location
50	Not specified	ARS = 65 mph[d]	DS = 60 mph	DS = 50-65 mph and not a high-accident location
Spot Vertical Curve Design Speed (mph)				
35	Not specified	ARS = 50 mph[d]	Crest DS = 45 mph Sag DS = 50 mph	DS = 35-50 mph, not a high-accident location, and no geometric features warranting special consideration (e.g., intersection)
40	Not specified	ARS = 55 mph[d]	Crest DS = 50 mph Sag DS = 55 mph	DS = 40-55 mph, not a high-accident location, and no geometric features warranting special consideration (e.g., intersection)
45	Not specified	ARS = 60 mph[d]	Crest DS = 55 mph Sag DS = 60 mph	DS = 45-60 mph
50	Not specified	ARS = 65 mph[d]	Crest DS = 60 mph Sag DS = 65 mph	DS = 50-65 mph

Mississippi	New Jersey[a]	South Dakota[b]	Texas	Washington
All highways	All highways for resurfacing and restoration projects	All highways	All highways	All highways
DS = 50 mph and not a high-accident location[d]	DS = 35 mph	Not specified (adequate for DS or signs should be provided)	Not specified (consider reconstruction if high-accident location)	DS = 50 mph and not a high-accident location[d]
DS = 55 mph and not a high-accident location[d]	DS = 40 mph	Not specified (adequate for DS or signs should be provided)	Not specified (consider reconstruction if high-accident location)	DS = 55 mph and not a high-accident location[d]
DS = 60 mph and not a high-accident location[d]	DS = 45 mph	Not specified (adequate for DS or signs should be provided)	Not specified (consider reconstruction if high-accident location)	DS = 60 mph and not a high-accident location[d]
DS = 65 mph and not a high-accident location[d]	DS = 50 mph	Not specified (adequate for DS or signs should be provided)	Not specified (consider reconstruction if high-accident location)	DS = 65 mph and not a high-accident location[d]
DS = 50 mph and not a high-accident location[d]	DS = 35 mph	Not specified (if 10 mph posted speed, signing required)	Not specified (consider reconstruction if high-accident location)	DS = 50 mph and not a high-accident location[d]
DS = 55 mph and not a high-accident location[d]	DS = 40 mph	Not specified (if 10 mph posted speed, signing required)	Not specified (consider reconstruction if high-accident location)	DS = 55 mph and not a high-accident location[d]
DS = 60 mph and not a high-accident location[d]	DS = 45 mph	Not specified (if 10 mph posted speed, signing required)	Not specified (consider reconstruction if high-accident location)	DS = 60 mph and not a high-accident location[d]
DS = 65 mph and not a high-accident location[d]	DS = 50 mph	Not specified (if 10 mph posted speed, signing required)	Not specified (consider reconstruction if high-accident location)	DS = 65 mph and not a high-accident location[d]

TABLE B-9 *continued*

Condition	California	Florida	Illinois	Michigan
Bridge Clear Width (ft)				
20	Not permitted	Not permitted	*Two-Lane Highways* ADT = 250	*Minor Rehabilitation* Where approach = 16 ft *Major Rehabilitation* ADT > 750 and where approach = 14 ft
22	Not permitted	Not permitted	*Two-Lane Highways* ADT = 400 DS = 50, ADT = 1,000	*Minor Rehabilitation* Where approach = 18 ft *Major Rehabilitation* ADT > 750 and where approach = 16 ft
24	ADT = 250	Where rehabilitated approach roadway = 20 ft	*Two-Lane Highways* ADT = 1,000 DS = 40, ADT = 3,000	*Minor Rehabilitation* Where approach = 20 ft *Major Rehabilitation* ADT > 750 and where approach = 18 ft
Stopping Sight Distance Less than AASHTO New Construction Standards (Horizontal Curves) [e]	When mitigated by regulations, signing, and so forth	When properly signed	Not specified	Not specified

NOTES: ARS = average running speed, DS = design speed, ADT = average daily traffic, DHV = design hourly volume, and T = percent trucks in traffic stream. Table entries are based on minimum values and specify the situations under which that stated condition may remain. In many cases, "not specified" means that although an exact number is not given in the state's standards, subjective treatment must be considered. Case study states fall into three general categories with regard to distinguishing between urban and rural standards: *(a)* states in which urban standards have not been developed (South Dakota); *(b)* states with separate urban provisions (Mississippi, Florida, Washington, Illinois, and Texas); and *(c)* states with essentially similar urban and rural standards (California, New Jersey, and Michigan). Most of the AASHTO new construction standards differentiate between urban and rural.

Mississippi	New Jersey[a]	South Dakota[b]	Texas	Washington
Minor Rehabilitation Permitted only if no accident problem exists and roadway is tapered before approach *Major Rehabilitation* Not permitted	Not permitted	Not permitted	*Two-Lane Highways* ADT = 400	ADT = 250 (except not permitted when rehabilitation is done on bridge)
Minor Rehabilitation Where approach = 22 ft *Major Rehabilitation* ADT = 750 or DHV = 200	*All Collectors* DHV = 200	Not permitted	*Two-Lane Highways* ADT = 400	ADT = 1,000 (except not permitted when rehabilitation is done on bridge)
Minor Rehabilitation Where approach = 24 ft *Major Rehabilitation* DHV = 400	*All Collectors* DHV = 400	All highways	*Two-Lane Highways* ADT = 750	ADT = 4,000 (except not permitted if rehabilitation is done on bridge)
When properly signed	Not permitted	When properly signed unless high-accident location	Not specified	When properly signed

[a] New Jersey distinguishes between resurfacing and restoration projects that are minor in nature and rehabilitation projects that involve more extensive safety-related improvements such as median barriers and minor widening.
[b] In South Dakota, "minor reconstruction" can involve minor widening, pavement rehabilitation, or shoulder improvements and thus are consistent with RRR definition.
[c] California's standards specify minimum "roadway" widths (lane plus shoulder) as follows:

ADT	Minimum Roadway, ft
3,000	24
3,000-5000	28
5,000	32

[d] In these states, if the ARS or DS is greater than that shown, consideration must be given to reconstructing the curve. If cost is prohibitive, additional signing is required.
[e] Stopping sight distance is also implicit in vertical curve design.

TABLE B-10 Comparison of Roadside Treatment Requirements in Special RRR Standards for Rural Highways, Case Study States

Condition	California	Florida	Illinois	Michigan
Clear zone (from edge of lane)	Not specified (removal of obstacles should be considered)	*ARS > 40 mph* 18 ft (if moved, 30 ft) *ARS <= 40 mph* 14 ft (if moved, 20 ft)	*Federal-Aid Secondary Routes* DS < 50 or ADT > 750: 10 ft or to ditch line Otherwise: 10 ft plus shoulder *Federal-Aid Primary/Federal-Aid Urban (uncurbed)* Posted speed >= 45: 18 ft Otherwise: 10 ft	Review accident history and potential for spot corrections
Maximum side slope[c]	Not specified (flattening of slopes should be considered)	Not specified	Not specified	*3:1* Two-lane, ADT < 750 Multilane undivided, ADT < 10,000 *4:1* All other highways
Culvert treatments	Not specified	Not specified	Extend to meet RRR roadway width If within clear zone, blend into slope If opening 30 in., use grates If opening 54 ft, protect with guardrail Review headwall positions	Not specified
Utilities	Not specified	May be retained if not a safety hazard; otherwise treat as under "clear zone"	Governed by state clear zone policy	Governed by state clear zone policy

Mississippi	New Jersey	South Dakota[a]	Texas	Washington
ARS > 40 Headwalls: 4 ft from shoulder Other: 14 ft If moved: 30 ft *ARS <= 40, ADT > 750* Headwalls: 2 ft from shoulder Other: 10 ft If moved: 20 ft *ARS <= 40, ADT < 750* Headwalls: 1 ft from shoulder Other: 6 ft If moved: 12 ft	Same as AASHTO new construction standards[b]	*Principal and Minor Arterials* Minor reconstruction: 25 ft if ADT < 1,000, 30 ft otherwise Resurfacing: 20 ft *Collectors* Minor reconstruction: 20 ft if ADT < 500, 25 ft otherwise Resurfacing: 10 ft (ADT < 250) 15 ft (250 < ADT < 500) 20 ft (ADT > 500)	*Rural Multilane Highways* 16 ft *Rural Two-Lane Highways* 7 ft, ADT < 750 16 ft, ADT >= 750	Not specified, but must be considered in a roadside hazard review report
Not specified	Same as AASHTO new construction standards[b]	If 3:1, slope must be justified, corrected, or protected	Retain existing sideslope except where grade or crown changes dictate otherwise	Do not steepen existing slopes; consider modifying if 3:1
Governed by state clear zone policy	Same as AASHTO new construction standards[c]	Not specified	*> 36 in.* If inside clear zone or creates a safety problem, treat by: 1. Grates, 2. Extending, or 3. Guard fence *<= 36 in.* If inside clear zone, replace with 3:1 or flatter culvert ends that blend into sideslope	Mitre end sections
Governed by state clear zone policy	If ROW available, locate outside clear zone unless cost is excessive or no accident problem If ROW not available, conduct analysis for relocation	Governed by state clear zone policy	Governed by state clear zone policy	Governed by state clear zone policy

TABLE B-10 *continued*

Condition	California	Florida	Illinois	Michigan
Other roadside provisions	None	Specimen trees and unique historical/environmental features, if a hazard, do not have to be removed if protected	Remove or upgrade guardrail Sign or light supports within clear zone should be breakaway Remove 4 in. or greater diameter trees or protect Remove 4 in. or higher concrete signal boxes within clear zone Where practical use impact attenuators instead of guardrail	Tree removal if frequent accident or target position of horizontal curve Obstructing sight distance at intersection Volunteer trees in clear zone Break existing treeline Retain if unique, scenic, or historic value

Mississippi	New Jersey	South Dakota[a]	Texas	Washington
Specimen trees and unique historical/ environmental features, if a hazard, do not have to be removed if protected Upgrade safety appurtenances (desirable)	All safety appurtenances should conform to state guidelines Roadside obstacles eliminated or shielded by longitudinal barriers	None	Upgrade or remove guardrail	Evaluate existing barriers and end treatments Smooth transition from guardrail to bridge rail Relocate, protect, or provide breakaway sign and lighting supports Protect bridge piers and abutments Modify raised drop inlets in clear zone

NOTE: ARS = average running speed, ADT = average daily traffic, DS = design speed, and ROW = right-of-way.

[a] In South Dakota, "minor reconstruction" is equivalent to minor widening and pavement rehabilitation.

[b] New Jersey's RRR standards state that AASHTO new construction standards are to be applied if a design feature is not explicitly mentioned in its standards.

[c] Refers to the region adjacent to the shoulder, more specifically known as the foreslope.

TABLE B-11 Federal-Aid Highway Projects, Case Study Cities

City	Population	Federal-Aid Category Received	Types of Projects Performed With Federal Aid
Columbus, Ohio	565,000	FAU	All types, including resurfacing, reconstruction, intersections, and bridges
San Antonio, Texas	785,000	FAU	Resurfacing, reconstruction, and intersections
Tallahassee, Florida	82,000	FAU	Mainly resurfacing and bridge rehabilitation because annual allotments are small
Orlando, Florida	130,000	FAU	Exclusively reconstruction due to capacity needs
Phoenix, Arizona	765,000	FAU	Exclusively reconstruction due to capacity needs
Cherry Hill, New Jersey	64,000	FAU	Exclusively reconstruction and intersections because of capacity and safety needs
Troy, New York	56,000	FAU	All types, including resurfacing, reconstruction, bridges, and intersections
Kansas City, Missouri	448,000	FAU, FA Bridge	Exclusively reconstruction (FAU) and bridge replacements and rehabilitations (FAU and FA Bridge)
Columbia, Missouri	62,000	FAU	Exclusively reconstruction and intersections (favors large projects to reduce number of federal-aid projects and associated federal grant procedures)
Moline, Illinois	46,000	FAU	Exclusively reconstruction, intersections, or safety hazards due to metropolitan planning organization technical ranking procedure (favors projects with big benefits)

TABLE B-11 *continued*

City	Population	Federal-Aid Category Received	Types of Projects Performed With Federal Aid
Decatur, Illinois	94,000	FAU, FA Bridge	Exclusively reconstruction (favors large projects to reduce number of federal-aid projects and associated federal grant procedures)
Rapid City, South Dakota	47,000	FAU	Almost exclusively reconstruction (one overlay project in last 5 years); favors reconstruction because of capacity needs
Sioux Falls, South Dakota	81,000	FAU	Exclusively reconstruction because of capacity needs
Lansing, Michigan	130,000	FAU, HES, FA Bridge	Traditionally reconstruction and intersections but shifting toward resurfacing and pavement rehabilitation. Michigan state law requires that 90 percent of all highway funds be spent on "preservation"; this also applies to federal aid.
Detroit, Michigan	1,203,000	FAU, HES, FA Bridge	All types, including resurfacing, reconstruction, intersections, and bridges
Madison, Wisconsin	171,000	FAU, HES	Exclusively reconstruction and intersections because of capacity needs
Knoxville, Tennessee	175,000	FAU, FA Bridge	Almost exclusively intersections because of local budget process, small annual amount of FAU funds, and restrictiveness of design standards

TABLE B-11 *continued*

City	Population	Federal-Aid Category Received	Types of Projects Performed With Federal Aid
Various cities in Mississippi	10,000–120,000	FAU	Traditionally intersections, but shifting toward RRR; favored because of small annual allotment of FAU funds. (In Mississippi it is common practice for one consulting engineer to manage under contract several cities' or counties' road programs. In the cases described here, one consulting engineer was responsible for several cities and another for several counties.)

NOTE: MPO—metropolitan planning organization, FAU—federal-aid urban fund, FA Bridge—federal-aid bridge replacement and rehabilitation funds, HES—federal-aid hazard elimination fund, FAS—federal-aid secondary fund, and RR crossing—federal-aid rail-highway grade crossing fund.

TABLE B-12 Federal-Aid Highway Projects, Case Study Counties

County	Population	Federal-Aid Category Received	Types of Projects Performed With Federal Aid
Franklin County, Ohio	100,000 (unincorporated)	FAU, FAS	Traditionally reconstruction but shifting toward RRR and intersections because of reduction in capacity needs
Dallas County, Texas	150,000 (unincorporated)	FAU	Exclusively reconstruction (at state's recommendation)
Four small counties in Mississippi	30,000–50,000	FAS	Exclusively reconstruction because of restrictiveness of standards
Coahoma County, Mississippi	37,000	FAS	Traditionally on bridges because of small allotment of annual FAS funds; shifting toward RRR
Maricopa County, Arizona	300,000 (unincorporated)	FAU, FAS	FAU: reconstruction because of capacity needs; some resurfacing. FAS: bridges because of safety needs; some resurfacing
Cape May County, New Jersey	82,000	FAS	Exclusively reconstruction and bridges because of restrictiveness of standards
Middlesex County, New Jersey	50,000 (unincorporated)	FAU, FAS	Exclusively reconstruction and bridges because of needs

TABLE B-12 *continued*

County	Population	Federal-Aid Category Received	Types of Projects Performed With Federal Aid
Schenectady County, New York	60,000 (unincorporated)	FAU, FAS	Exclusively reconstruction and bridges because of capacity and safety needs plus impression that state favors larger projects
St. Louis County, Missouri	200,000 (unincorporated)	FAU	Traditionally reconstruction because of capacity needs but resurfacing and rehabilitation projects are now programmed
Union County, South Dakota	11,000	FAS, FA Bridge	Exclusively resurfacing and bridges because of small annual allotment of federal funds
Ingham County, Michigan	90,000 (unincorporated)	FAS, FAU, HES, RR Crossing	Mostly resurfacing and rehabilitation because of state's 90 percent preservation requirement
Los Angeles County, California	7,477,000 (total)	FAU	Traditionally reconstruction because of capacity needs but now all types including reconstruction, RRR, intersections and bridges
Shasta County, California	115,000 (total)	FAS, FAU	Mostly resurfacing because of need; some reconstruction.

NOTE: MPO—metropolitan planning organization, FAU—federal-aid urban fund, FA bridge—federal-aid bridge replacement and rehabilitation funds, HES—federal-aid hazard elimination fund, FAS—federal-aid secondary fund, and RR crossing—federal-aid rail-highway grade crossing fund.

TABLE B-13 Federal-Aid Highway Projects, Case-Study Metropolitan Planning Organizations

MPOs	Population	Federal-Aid Category Received	Types of Projects Performed With Federal Aid
Capital District Transportation Authority, Albany, New York	430,000 (urbanized area)	FAU	Mostly intersections, traffic operations, and bridges but also reconstruction and RRR-type minor widenings (no resurfacing); project selection primarily because of MPO technical ranking procedure
Erie County, New York Planning Commission	850,000 (urbanized area)	FAU	Exclusively reconstruction primarily because of capacity needs but also past problems with standards capacity
Chicago Area Transportation Study	6,000,000 (approximate)	FAU	Traditionally reconstruction and new construction but shifting now to a balance among all types

NOTE: MPO—metropolitan planning organization, FAU—federal-aid urban fund, FA bridge—federal-aid bridge replacement and rehabilitation funds, HES—federal-aid hazard elimination fund, FAS—federal-aid secondary fund, and RR crossing—federal-aid rail-highway grade crossing fund.

TABLE B-14 Allocation of Federal Aid to Local Governments, Case Study States

State Highway Agency Practices	Ariz.	Calif.	Fla.	Ill.	Mich.	Miss.	Mo.	N.H.	N.J.	N.Y.	Ohio	S.Dak.	Tex.	Va.	Wash.
Allocates funds to local governments by formula															
Attributable urban system	Yes	Yes	Yes	Yes	Yes	Yes	Yes	Yes	Yes	Yes	Yes	NA	Yes	Yes	Yes
Nonattributable urban system		Yes	No	Yes[a]	Yes[a]	Yes	Yes	Yes	Yes	No	Yes[a]	Yes	No	Yes	Yes
Secondary	Yes	Yes	No	Yes	Yes	Yes	NA[b]	NA[b]	No	No	Yes	Yes	NA[b]	NA[b]	Yes
Provides matching funds for local federal-aid projects															
Urban system	No	No	Yes	No	No	No	No[c]	No	Yes	Yes[d]	Yes	Yes	Yes	Yes	No
Secondary	No	No	Yes	No	No	No	NA[b]	NA[b]	Yes	Yes	Yes	No	NA[b]	NA[b]	No
Provides design and construction assistance for local federal-aid projects															
Performs most design	No	No	Yes	No	No	No	No	Yes	Yes	Yes	No	No	Yes	Yes	No
Lets most construction contracts	No	No	Yes	Yes	Yes	No	No	Yes	Yes	Yes	Yes	Yes	Yes	Yes	No

[a] Some Nonattributable funds are retained by state and dispensed to smaller cities on a discretionary basis.
[b] Not applicable because nearly all secondary highways are state maintained; state agency may still allocate funds by formula among its administrative substate areas.
[c] If the federal-aid urban highway is also a state route, the state provides the full match.
[d] Eighty percent of matching share is provided by state.

TABLE B-15 Design Standards Used for Local Federal-Aid RRR Projects, Case Study States

State	Has State Developed Special Federal-Aid RRR Standards?	Standards Applied to Local Federal RRR Projects	
		Urban System	Secondary System
Arizona	No	AASHTO new construction	AASHTO new construction
California	Yes	State's RRR standards[a]	State's RRR standards
Florida	Yes	State's RRR standards	State's RRR standards
Illinois	Yes	Either state's RRR standards or AASHTO new construction	Either state's RRR standards or AASHTO new construction
Michigan	Yes	Special local RRR standards[b]	Special local RRR standards[b]
Mississippi	Yes	State's RRR standards[c]	State's RRR standards
Missouri	No	AASHTO new construction	NA[d]
New Hampshire	No	AASHTO new construction	NA[d]
New Jersey	Yes	State's RRR standards	State's RRR standards
New York	No	AASHTO new construction	AASHTO new construction
Ohio	No	AASHTO new construction	AASHTO new construction
South Dakota	Yes	AASHTO new construction	Special RRR standards specified in secondary road plan
Texas	Yes	State's RRR standards	NA[d]
Virginia	No	AASHTO new construction	AASHTO new construction
Washington	Yes	Special local RRR standards[b]	Special local RRR standards[b]

[a]Los Angeles County has developed its own RRR standards for federal-aid projects.
[b]Standards that are distinct from the ones applied to state-administered projects.
[c]Arterials only; AASHTO new construction standards apply to urban collectors and local streets.
[d]Not applicable because 90 percent or more of secondary highways are state maintained.

Appendix C
Summary of Detailed Safety Relationships

As a part of this study, detailed relationships were developed that describe the likely effects of the following design features on highway accidents: *(a)* lane and shoulder conditions, *(b)* bridge width, *(c)* horizontal and vertical curvature, and *(d)* roadside obstacles. The study focused on two-lane rural highways of the type eligible for resurfacing, restoration, and rehabilitation (RRR) funding.

Considerable judgment was required in developing these relationships. Although they have been useful in several phases of the study, none has been validated in a critical way. Until such testing has been completed and appropriate modifications made, the relationships must be considered as only representing reasonable, most likely safety effects of these roadway and roadside elements. Their most promising interim use is to estimate the accident reductions likely to result from incremental roadway and roadside improvements.

The purpose of this appendix is to summarize these safety relationships and to present a method for estimating the combined effects of simultaneous improvements.

LANE AND SHOULDER WIDTH AND SHOULDER TYPE

Lane and shoulder conditions have been found to influence the frequency of accidents but not necessarily accident severity. Safety is enhanced by increases in lane and shoulder widths and improvements in shoulder surface type.

The safety effect of lane and shoulder width and shoulder type can be estimated as follows *(1)*:

$$A = 0.0019 \, (ADT)^{0.882} \, (0.879)^{W} \, (0.919)^{PA} \, (0.932)^{UP}$$
$$(1.236)^{H} \, (0.882)^{TER1} \, (1.322)^{TER2} \qquad (1)$$

where

- A = number of run-off-road, head on, opposite-direction sideswipe, and same-direction sideswipe accidents per mile per year;
- ADT = two-directional average daily traffic volume;
- W = lane width in feet;
- PA = width of paved shoulder in feet;
- UP = width of unpaved (gravel, turf, earth) shoulder in feet;
- H = median roadside hazard rating for the highway segment, measured subjectively on a scale from 1 (least hazardous) to 7 (most hazardous);
- $TER1$ = 1 for flat terrain, 0 otherwise; and
- $TER2$ = 1 for mountainous terrain, 0 otherwise.

This accident model is limited in that it applies to

- Lane widths of 8 to 12 ft and shoulder widths of 0 to 10 ft. Combinations of lane and shoulder widths that can be reasonably modeled are limited to those shown in Figure 3-2, Chapter 3;
- Two-lane, two-way paved rural roads on state primary and secondary systems; and
- Homogeneous roadway sections, and does not include the additional accidents expected at intersections.

BRIDGE WIDTH

Although the hazard associated with narrow bridges has been recognized for many years, efforts to quantitatively establish the influence of bridge cross section and geometry on accident frequency and severity have realized limited success. The more acceptable of these efforts use relative bridge width as the appropriate physical measure. Relative bridge width is defined as the difference between the clear bridge width, including both traffic lanes and usable shoulders, and the total width of the traffic lanes, excluding the shoulders, on the approach to the bridge.

The rate of bridge-related accidents on two-lane highways can be estimated as follows *(2)*:

$$AR = 0.50 - 0.061\ (RW) + 0.0022(RW)^2 \quad \text{for } 0 \leq RW \leq 14 \tag{2}$$

where AR is the number of accidents per million vehicles and RW is the relative bridge width in feet. Equation 2 does not apply in situations where the width of the approach traffic lanes exceeds the clear bridge width: in this region, the accident rate is greatly increased by further constriction in the traffic lanes on the bridge. Nor does it apply for relative bridge widths in excess of about 14 ft, a region where the upturn in computed accident rates is more likely an artifact of the model-building process than a valid indication of impaired safety. Note that the measure of exposure to narrow-bridge hazard, as expressed in Equation 2, is the total number of vehicle traversals and not the more commonly used vehicle miles of travel.

Factors other than relative bridge width, such as bridge length and type (e.g., deck versus truss), the presence or absence of curbs, approach alignment, pavement surface condition, and so forth, may also affect the accident rate at bridges. These factors are not included in Equation 2 because definitive information on their safety effects is unavailable in the published literature. No evidence exists to suggest a relationship between the severity of constriction at bridges and the severity of bridge-related accidents.

HORIZONTAL CURVATURE

Accidents are more likely to occur on horizontal curves than on straight, or tangent segments of roadway because of increased demands placed on the driver, the vehicle, and friction at the tire-pavement interface. Accident frequency on a segment of roadway containing a single horizontal curve and its tangent approaches can be estimated as follows (Appendix D):

$$A = AR_s\ (L)\ (V) + 0.0336\ (D)\ (V) \quad \text{for } L \geq L_c \tag{3}$$

where

A = total number of accidents on the segment,
AR_s = accident rate on comparable straight segments in accidents per million vehicle miles,
L = length of highway segment in miles,
V = traffic volume in millions of vehicles,
D = curvature in degrees, and
L_c = length of curved component in miles.

The accuracy of Equation 3 may be diminished for curves sharper than about 15 degrees, the approximate limit recorded in the data base from which the model was calibrated.

Similar to the most likely relationship for accidents at narrow bridges (Equation 2), the total number of vehicle traversals rather than vehicle miles of travel is used to represent exposure to the extra hazards of travel at curve locations. Equation 3 does not capture the effects of other physical features of curves—superelevation and superelevation runoff, cross-slope break, spiral transitioning, curve length, and so forth—that likely have significant, but unquantified, effects on accidents. Neither does it reflect the effects of roadway uniformity on driver expectations: a sharp curve immediately following an extended stretch of straight highway will experience more accidents than a similar curve situated within a generally winding section.

Possible effects of horizontal curvature on accident severity have not been identified in the literature. The propensity for off-roadway encroachments onto the outside of horizontal curves has been well established, however, and it is clear that roadside conditions in this region can substantially affect accident severity and, quite possibly, accident frequency as well.

VERTICAL CURVATURE

The primary effect of vertical curvature on highway safety is related to possible restrictions on sight distance that adversely affect emergency avoidance maneuvers. Both the severity and length of the sight-distance restriction are significant determinants of accident frequency. Accident frequency on a segment of roadway containing a single crest vertical curve and its tangent approaches can be estimated as

$$N = AR_h (L) (V) + AR_h (L_r) (V) (F_{ar}) \qquad (4)$$

where

N = number of accidents on a segment of highway containing a crest curve,
AR_h = average accident rate for the specific highway—or alternatively for the related general highway class—in accidents per million vehicle miles,
L = length of highway segment in miles,
V = traffic volume in millions of vehicles,
L_r = length of restricted sight distance in miles, and
F_{ar} = a hypothetical accident rate factor that varies according to both

the severity of the sight restriction and the nature of the hidden hazard.

Procedures for estimating L_r and for selecting F_{ar} are detailed in Appendix E.

Because Equation 4 has not been validated by comparison with actual accident experience, it clearly must be used with considerable caution. Logically, it does appear to incorporate the primary effects of restricted lines of sight, although only superficial treatment is given to characterizations of the severity of the hidden hazard. With the possible exception of drainage on roadways with curbs, little or no evidence exists to suggest that factors other than restricted lines of sight affect the safety of operations on vertical curves. The geometry of vertical curves is not known to have a significant effect on accident severity.

ROADSIDE FEATURES

The frequency of roadside accidents can be reduced by geometric improvements such as lane widening and curve lengthening and by traffic control measures such as edge lines and post-mounted delineators, all of which are designed in part to reduce the frequency of inadvertent roadside encroachments. Perhaps the most beneficial roadside improvement is the provision of a clear zone adjacent to the travel lanes, an obstacle-free area of traversable slope. The likelihood of a run-off-road accident diminishes rapidly as potential hazards are further displaced from the roadway: even relatively narrow recovery areas are remarkably effective. Recent developments in the design of safer drainage systems provide important opportunities for enhancing safety on RRR projects *(3)*. Additional roadside improvements include flattening of sideslopes to reduce the likelihood of overturn, and redesign of roadside hardware to reduce the severity of impact. When necessary, the danger of a hazardous roadside condition is lessened by the provision of guardrail to contain and redirect the errant vehicle.

The impact of the roadside environment on highway safety has proven to be difficult to establish. Useful composite measures of the degree of roadside hazard have been elusive, and the diversity of the many potential hazards has been almost overwhelming. However, a review commissioned by this study in conjunction with FHWA *(1)* revealed a significant relationship between roadside recovery distance (measured from the outside edge of the shoulder to the nearest roadside obstacles or hazards) and the number of single vehicle, head-on, and sideswipe accidents on two-lane rural highways:

Roadside Recovery Distance (ft)	Percent Reduction In Accidents
5	13
10	25
15	35
20	44

Roadside encroachment models are often used to examine the safety effects of specific roadside features. The following model is recommended for such use:

$$Ex(A_h) = \frac{0.07285(ADT)^{0.5935}}{21,120} Pr(A_h|C_h)(\sum_i x_i e^{-0.08224 y_i} + \sum_j x_j e^{-0.08224 y_j}) \tag{5}$$

where

$Ex(A_h)$ = expected annual number of accidents involving the hazard;
ADT = two-directional average daily traffic volume;
$Pr(A_h|C_h)$ = probability that an encounter or collision with the hazard will result in an accident (Table F-1, Appendix F); and
x's and y's = projected length along the travel lanes and offset to elements of the hazard, respectively.

The x's and y's reflect anticipated encroachment angles of errant vehicles, as well as the offset to and dimensions of the roadside hazard and are chosen in part—following the example of Figure F-1 and Table F-2, Appendix F—for computational convenience. The subscript, i, in Equation 5 pertains to near-side encroachments whereas the subscript, j, refers to far-side encroachments.

Accident reduction factors, computed using Equation 5, are applicable only to the run-off-road portion of the total accident population. They are likely to understate the reduction achievable at particularly critical locations such as the outside of sharp, horizontal curves. In addition to the number of accidents, the severity of run-off road accidents can also be estimated (Table F-1, Appendix F).

COMBINED EFFECTS

The task of sorting out possible safety effects of incremental roadway and roadside improvements is best performed through the application of accident reduction factors to historical accident data collected at the location being improved. Should actual accident data be unavailable, however, the analyst must use models, calibrated from other accident-data sources, that directly estimate either accidents or accident rates. These models are not expected to be as reliable as the use of accident reduction factors because of their additional abstraction from the location in question.

For either type of analysis, the major question involves the future; that is, how the accident pattern is likely to differ if the improvement is or is not made. When the likely change in accidents is estimated by applying accident reduction factors to the historical accident record, care must be taken to assure that anticipated changes in traffic volume are properly accounted for.

More difficult to treat than the volume influence is the estimation of accident effects for situations in which two or more highway features are improved simultaneously. Interactive effects of such combined treatments can result in any of three possible consequences: improvement to one highway feature

- *Diminishes* the relative safety effect of an improvement to another feature;
- *Has no influence* on the relative safety effect of improvement to another feature; or
- *Enhances* the relative safety effect of improvement to another feature.

Because of difficulty both in experimental design and model calibration *(4)*, few interactive accident models are available. Of the models developed for use here, none incorporates interactive effects. In the absence of interactive models, ad-hoc procedures are required to estimate the accident reduction likely to result from two or more simultaneous improvements.

A first-order approximation of combined effects, in use for many years *(5)*, is illustrated for three simultaneous improvements as follows:

$$\frac{ARF_c}{100} = 1 - (1 - \frac{ARF_1}{100})(1 - \frac{ARF_2}{100})(1 - \frac{ARF_3}{100}) \qquad (6)$$

where ARF_c is the accident reduction factor, expressed as a percentage, expected from the combination of improvements, and ARF_1, ARF_2, and ARF_3 are similar factors representing the effects of the first, second, and third improvements, respectively.

Although use of Equation 6 is recommended for estimating combined effects, it is necessary to emphasize that it does not account for interactions among the combined improvements *(6)*. Consequences of this deficiency will be ameliorated by

- Categorizing the accident pattern into subsets, consonant with the types of accidents likely to be affected by the improvements being made;
- Applying Equation 6 only for that subset(s) jointly influenced by the combined improvements; and
- Applying a single accident reduction factor—usually the more significant one—in situations where one improvement is likely to substantially negate influences of another. [Such a situation is most likely when joint improvements are intended to affect the same subset of the accident population by similar means. A good example is the addition of both centerline and edge striping together with post-mounted delineators at a hazardous horizontal curve. In this example, each mitigation measure is designed to improve delineation.]

Also, a minimum estimated post-improvement accident rate—based on accident experience for the best designed and operated facilities—is useful to guard against overestimating the safety benefits of improvements *(7)*.

REFERENCES

1. C. V. Zegeer, J. Hummer, D. Reinfurt, L. Hurf, and W. Hunter. *Safety Effects of Cross-Section Design for Two-Lane Roads,* Vols. I and II. Report FHWA-RD-87/008 and 009. FHWA, U.S. Department of Transportation, 1986.
2. D. S. Turner. "Prediction of Bridge Accident Rates." *Journal of Transportation Engineering,* Vol. 110, No. 1, American Society of Civil Engineers, New York, Jan. 1984.
3. H. D. Robertson, S. Basu, K. Colpitts, S. Stein, F. Johnson, and G. K. Young. *Safer Drainage Systems.* FHWA Office of Research and Development, U.S. Department of Transportation, July 1986.
4. F. M. Council et al. *Accident Research Manual.* Report FHWA-RD-80/016. FHWA, U.S. Department of Transportation, Feb. 1980.
5. Roy Jorgensen Associates. *Evaluation of Criteria for Safety Improvements on the Highway.* Gaithersburg, Md., 1966.
6. H. D. Robertson and K. S. Opiela. *Feasibility of Determining the Incremental Effectiveness of Accident Countermeasures.* Report FHWA-RD-85/043. FHWA, U.S. Department of Transportation, Feb. 1985.
7. T. N. Tamburri and R. N. Smith. "The Safety Index: A Method of Evaluating and Rating Safety Benefits." In *Highway Research Record 332,* TRB, National Research Council, Washington, D.C., 1970, pp. 28–39.

Appendix D
Relationship Between Accidents and Horizontal Curvature

Although a number of researchers have suggested a relationship between accident rates at horizontal curves and the radius or degree of curvature, the validity of these relationships is often unknown because of questionable experimental design and the imprecise definition of both curve-related accidents and vehicle exposure *(1)*. Among the research studies currently available, a recent study by Glennon et al. *(2)*, sponsored by the Federal Highway Administration, is one of the best. In this study proper experimental design was applied in assembling a data base that includes accident, geometric, and traffic data for two-lane rural highway segments, each of which included a horizontal curve. The relationship between accidents and horizontal curvature reported in Chapter 3 and used in Chapter 5 for cost-effectiveness analyses of curve flattening projects is based on these data and the statistical analyses reported by Glennon et al. For site-specific design, the relationship provides a helpful estimate of the accident reduction that may be possible by flattening a curve, but it must be used judiciously along with other pertinent information such as traffic characteristics, adjacent highway alignment, shoulder and roadside characteristics, and prior accident experience. A summary of the data set, and the most likely quantitative relationship is presented next.

GLENNON, NEUMAN, AND LEISCH DATA SET

Highway sites (or segments) in this data base were typically 0.61 mi long or longer as necessary to assure straight (tangent) components at least 650 ft long at each end of a horizontal curve. Care was taken to select sites with uniform lane and shoulder conditions and to avoid influences of major bridges and intersections, curbs, and other nearby horizontal curves. Furthermore, selection was limited to sites without reconstruction or significant traffic growth during the 3-year study period. Sites were located in four states (Florida, Illinois, Ohio, and Texas) and limited to two-lane rural roads with daily traffic averaging 1,500 or more vehicles. The limited number of straight, control sites was comparable to the curved sites both in traffic volume and in lane and shoulder conditions (Table D-1).

TABLE D-1 Summary Statistics of Glennon et al. Data Base *(2)*

	Type of Site	
	Straight	Curved
Number of sites	351	3,297
Average ADT (VPD)	3,417	3,184
Average site length (mi)	0.632	0.631
Average No. of accidents (3 years)	2.22	3.93
Average accident rate (PMVM)	0.902	1.823
Average lane width (ft)	11.50	11.45
Average shoulder width (ft)	7.20	6.87
Average degree of curvature	–	3.47
Average curve length (mi)	–	0.171

NOTES: ADT = average daily traffic; VPD = vehicle per day; PMVM = per million vehicle miles. Dashes (–) indicate not applicable.

In terms of safety, the average curved site was distinctly inferior to the average straight site (Table D-1). An approximate proration of the accidents on the average curved site to its straight and curved components indicates that the accident rate on the curve section was three to four times the rate on the straight sections.

Except for what may be an anomaly at the stratification for the smallest degree of curvature (which persists in later tabulations as well), the average accident rate significantly increased with increased degree of curvature (Table D-2). Even larger increases would be expected had it been possible to separate the effects of the curved and straight elements within each of the nominally 0.61-mi sites.

It is likely that accident rates, such as those reported in Table D-2, are biased by unknown influences of other safety-related variables. It is reasonable to expect, for example, that sites containing the sharper curves had geometric and roadside conditions that were also likely to have been more hazardous. Although this possibility is generally confirmed by the trends in average lane and shoulder widths as reported in Table D-2, the differences do

TABLE D-2 Average Accident Rate (PMVM) on Nominally 0.61-mi Site as a Function of Degree of Curvature

Degree of Curve	Average Accident Rate (PMVM)	Average ADT (VPD)	Average Lane Width (ft)	Average Shoulder Width (ft)
0	0.90	3,400	11.5	7.2
0.01–0.74	1.38	3,100	11.7	7.7
0.75–1.49	1.06	3,300	11.9	7.5
1.50–2.49	1.24	3,200	11.8	7.4
2.50–3.49	1.61	3,400	11.7	7.3
3.50–4.49	2.41	3,000	11.3	6.3
4.50–6.49	2.79	3,200	10.9	5.9
6.50–8.49	2.89	3,300	10.4	4.8
8.50–10.49	3.59	3,000	10.2	4.8
10.50–12.49	4.03	3,200	10.3	4.8
12.50 or more	4.19	2,900	10.0	4.8

NOTE: Tabulations are based on data from Glennon et al. data base *(2)*.

not appear as large as might have otherwise been anticipated. Based on the analysis by Zegeer and Deacon *(3)*, the reduced lane and shoulder widths for the sharper curves could be expected to account for only about 10 percent of the difference in the accident rates between the sharpest and flattest sites. Additional cross tabulations indicate that lane width may have had a minor effect on reported accident rates (Table D-3) and that volume effects are likely to have been small (Table D-4). In addition, once degree of curvature is taken into account, the data show no consistent and pronounced relationship

TABLE D-3 Average Accident Rate (PMVM) on Nominally 0.61-mi Site as a Function of Degree of Curvature and Lane Width

| Degree of Curve | Lane Width (ft) | | | |
	9 or Less	10	11	12 or More
0	*0.96*	*1.17*	0.90	*0.78*
0.01–0.74	1.69	*1.21*	1.32	*1.40*
0.75–1.49	1.54	*1.01*	1.35	*0.98*
1.50–2.49	1.76	1.30	*1.30*	1.18
2.50–3.49	1.95	1.67	1.93	*1.52*
3.50–4.49	2.64	*2.45*	2.49	*2.24*
4.50–6.49	3.08	*2.92*	2.82	*2.61*
6.50–8.49	3.65	2.69	5.29	2.11
8.50–10.49	*3.90*	3.44	2.24	*3.82*
10.50–12.49	4.34	*4.11*	4.57	3.66
12.50 or more	*4.15*	4.76	2.08	2.69

NOTES: Italicized numbers indicate 30 or more sites. Tabulations are based on data from Glennon et al. data base *(2)*.

TABLE D-4 Average Accident Rate (PMVM) on Nominally 0.61-mi Site as a Function of Degree of Curvature and Volume

Degree of Curve	Volume (VPD)			
	2,099 or less	2,100–3,099	3,100–4,899	4,900 or more
0	0.85	0.74	0.72	1.55
0.01–0.74	1.93	1.10	1.00	1.32
0.75–1.49	1.06	1.08	1.05	1.01
1.50–2.49	1.31	1.21	1.22	1.14
2.50–3.49	1.85	1.49	1.71	1.26
3.50–4.49	2.36	2.12	2.73	2.81
4.50–6.49	2.90	2.91	2.49	2.77
6.50–8.49	2.68	2.61	3.46	2.97
8.50–10.49	3.73	3.56	3.51	3.47
10.50–12.49	3.64	4.71	3.74	4.45
12.50 or more	4.15	3.94	3.91	6.17

NOTES: Italicized numbers indicate 30 or more sites. Tabulations are based on data from Glennon et al. data base *(2)*.

between accident rate and either curve length or curve central angle, a measure of the total change in direction of the highway (Tables D-5 and D-6).

DEVELOPMENT OF ACCIDENT MODEL

Concerned with the possible confounding effects of other variables (state, length of curve, lane width, and shoulder width), Glennon et al. *(2)* used analysis-of-covariance techniques to isolate the incremental effects of changes in degree (or sharpness) of curvature on accidents. The estimated reduction in the number of accidents per million vehicles for a one-degree change in curvature was found to be 0.0336, a number that served as the basis for development of a complete accident model.

To develop the model, the number of accidents at the combined straight-curved sites, A, can be represented as the sum of two components:

$$A = A_s + A_c$$

where A is the number of accidents on the straight segments and A_c is the number of accidents on the curved segment. If AR_s is the accident rate (PMVM) on the straight roadway, then

$$A = AR_s\ (L_s)(V) + A_c$$

or

$$A = AR_s\ (L)(V) + [A_c - AR_s\ (L_c)(V)]$$

and

$$A = AR_s(L)(V) + \Delta A_c \tag{1}$$

where

L = length of site in miles,
L_s = length of straight segments in miles,
L_c = length of curved segment in miles, and
V = traffic volume in millions.

ΔA_c, the increase in accidents as a result of curvature beyond that anticipated for a straight roadway of the same length, was calibrated as follows:

$$\Delta A_c = 0.0336(D)(V)$$

where D is the degree of curvature.

The complete accident model can then be expressed as

$$A = AR_s(L)(V) + 0.0336(D)(V) \quad \text{for } L \geq L_c \tag{2}$$

and, on the curved segment alone,

$$A_c = AR_s(L_c)(V) + 0.0336(D)(V) \tag{3}$$

From the data base, $AR_s = 0.902$ (Table D-1).

Equation 3 has two components: the first represents a steady-state turning effect and the second represents transitional (entry and exit) effects. The steady-state-turning component is directly proportional to the vehicle miles of travel on the curve but is insensitive to degree of curvature. The transitional component is directly proportional to both degree of curvature and traffic volume. Other model formulations are possible, and a number were calibrated as part of this study. It was concluded, however, that significant improvements to the Glennon et al. model (using the existing data base) are unlikely and that this model (Equations 2 and 3) represents the most acceptable technique currently available for estimating the safety effects of curve flattening.

MODEL APPLICATION

The model described was used to analyze the safety-cost effectiveness of different RRR policies for improvement of horizontal curves (Chapter 5). As

TABLE D-6 Average Accident Rate (PMVM) on Nominally 0.61-mi Site as a Function of Degree of Curvature and Central Angle

Degree of Curve	Central Angle (deg.)										
	2.49 or less	2.50–7.49	7.50–12.49	12.50–17.49	17.50–22.49	22.50–27.49	27.50–32.49	32.50–37.49	37.50–42.49	42.50–47.49	47.50–52.49	52.5 or More
0.01–0.74	*1.28*	*1.24*	*1.40*	*1.47*	*1.08*	*1.45*	*1.05*	3.42	1.92	–	–	–
0.75–1.49	*1.06*	*1.12*	*0.96*	*0.98*	*1.26*	*1.21*	*0.92*	0.78	1.35	0.72	2.42	0.71
1.50–2.49	*4.67*	*1.05*	*1.39*	*1.25*	*1.13*	*1.22*	*1.22*	0.95	1.26	1.00	1.54	0.93
2.50–3.49	0.00	2.70	1.20	*1.54*	*1.14*	*1.46*	*1.75*	*1.49*	1.35	2.45	1.53	1.04
3.50–4.49	4.09	3.61	3.38	2.41	2.42	1.83	2.75	2.51	2.02	2.42	3.36	2.72
4.50–6.49	0.00	1.74	3.03	3.00	2.87	2.67	2.56	3.04	2.92	1.79	2.20	2.61
6.50–8.49	–	2.45	2.98	2.87	3.92	2.84	2.85	2.76	3.47	3.69	2.91	2.12
8.50–10.49	–	–	3.24	3.98	3.35	3.92	4.06	3.05	4.21	3.30	–	3.76
10.50–12.49	–	2.11	5.04	3.83	–	5.19	3.83	3.96	4.55	3.28	2.06	2.51
12.50 or more	–	–	5.33	2.58	3.74	4.82	6.30	4.70	4.18	2.66	3.62	4.37
All	*1.30*	*1.24*	*1.61*	*1.83*	*1.87*	2.03	2.37	2.33	2.44	2.04	2.37	2.74

NOTES: Italicized numbers indicate 30 or more sites; dashes (–) indicate no sites. Tabulations are based on data from Glennon et al. data base *(2)*.

TABLE D-5 Average Accident Rate (PMVM) on Nominally 0.61-mi Site as a Function of Degree and Length of Curvature

Degree of Curve	Length (mi)								
	0.000–0.049	0.050–0.069	0.070–0.089	0.090–0.109	0.110–0.139	0.140–0.169	0.170–0.209	0.210–0.259	0.260–0.319	0.320 or more
0.01–0.74	1.78	1.19	1.60	1.01	1.57	1.65	1.41	1.15	0.87	1.19
0.75–1.49	1.24	1.03	0.91	1.41	1.04	0.98	1.14	0.94	0.83	1.16
1.50–2.49	1.43	1.13	1.65	0.95	1.85	0.98	1.32	1.16	1.05	1.14
2.50–3.49	2.90	0.92	1.57	1.40	1.34	1.73	1.65	1.36	2.48	1.03
3.50–4.49	3.54	2.67	2.36	2.17	1.77	2.59	2.05	2.74	2.74	2.44
4.50–6.49	2.99	3.24	2.53	2.40	3.19	2.54	2.33	1.62	2.20	2.89
6.50–8.49	2.85	3.24	3.02	2.01	3.56	2.02	1.71	3.49	0.00	2.36
8.50–10.49	3.46	4.06	3.59	2.43	4.57	4.14	2.70	–	–	–
10.50–12.49	4.80	3.86	3.58	2.19	2.25	3.41	–	–	–	–
12.50 or more	4.63	4.43	3.56	3.24	2.55	3.65	1.09	1.26	–	–
All	2.32	2.74	2.10	1.49	1.79	1.59	1.47	1.24	1.19	1.20

NOTES: Italicized numbers indicate 30 or more sites; dashes (–) indicate no sites. Tabulations are based on data from Glennon et al. data base (2).

noted earlier, this relationship should not be applied to specific project design situations without careful consideration of a variety of site-specific factors. When it is used, the coefficient, 0.902, should be replaced where possible by an accident rate on straight segments representative of local conditions for the highway under consideration.

Equation 2 can be used to estimate the reduction in accidents from flattening a horizontal curve while maintaining its lines of tangency (or central angle). Because curve flattening reduces the overall length of highway—as well as increases the length of its curved component—before-and-after comparisons must be made between two points on the original alignment whose locations remain unchanged by the improvement. Any two points, each separated from the point of intersection by at least the tangent distance of the new alignment, will be sufficient.

Using Equation 2, the net reduction in the number of accidents, ΔA, is as follows:

$$\Delta A = A_o - A_n$$

$$\Delta A = AR_s (L_o - L_n)V + 0.0336 (D_o - D_n)V$$

$$\Delta A = AR_s(\Delta L)V + 0.0336 (\Delta D)V \tag{4}$$

where the subscript, o, refers to the original alignment and the subscript, n, the new or improved alignment. The amount by which the alignment length is shortened, ΔL, can be determined as

$$\Delta L = [(2.170 \tan I/2) - (I/52.8)] [(1/D_n) - (1/D_o)] \tag{5}$$

where ΔL is expressed in miles and I is the central angle in degrees. At central angles of 90 degrees and less, ΔL is sufficiently small that it can be ignored.

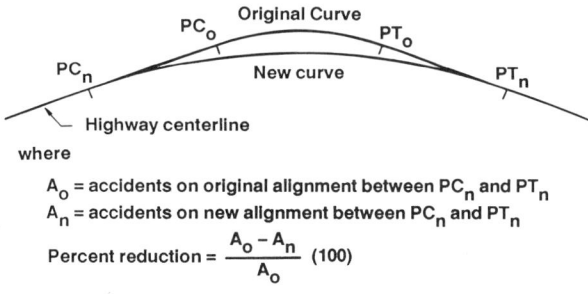

FIGURE D-1 Basis for calculating accident reduction percentiles reported in Table D-7.

TABLE D-7 Percentage Reduction in Accidents Due to Horizontal Curve Flattening

Original Degree of Curve	New Degree of Curve	Central Angle (deg.)				
		50	40	30	20	10
30	25	16	16	16	16	16
	20	32	33	33	33	33
	15	49	49	49	49	50
	10	65	65	66	66	66
	5	82	82	82	82	83
25	20	19	19	20	20	20
	15	39	39	39	39	40
	10	58	58	59	59	60
	5	78	78	78	79	79
20	15	24	24	24	24	25
	10	47	48	48	49	49
	5	72	72	72	73	74
15	10	30	31	31	32	33
	5	61	62	63	64	65
10	5	41	42	44	45	48

NOTE: Based on $AR_s = 0.902$ accidents PMVM, the average accident rate at straight sites from Table D-1. Percentages based on the number of accidents with original curve between the points of tangency of the new curve (see Figure D-1).

For a given change in the degree of curvature, the greatest accident reductions—when expressed as a percentage of the number of accidents on the original alignment, which occur between the tangent points of the new curve (Figure D-1)—are expected for the smallest central angles and the sharpest original curvature (Table D-7). For given initial conditions (D_o and I), accidents decrease with reductions in the degree of curvature (D_n).

REFERENCES

1. J. C. Glennon. "Effect of Alinement on Highway Safety: A Synthesis of Prior Research." In TRB State-of-the-Art Report. TRB, National Research Council, Washington, D.C. (forthcoming).
2. J. C. Glennon, T. R. Neuman and J. E. Leisch. *Safety and Operational Considerations for Design of Rural Highway Curves*. Report FHWA-RD-86/035. FHWA, U.S. Department of Transportation, Aug. 1983.
3. C. V. Zegeer and J. A. Deacon. "Effect of Lane Width, Shoulder Width, and Shoulder Type on Highway Safety: A Synthesis of Prior Research." In TRB State-of-the-Art Report. TRB, National Research Council, Washington, D.C. (forthcoming).

Appendix E
Relationship Between Accidents and Sight Distance at Crest Vertical Curves

Neuman and Glennon have developed a hypothetical model for estimating the effects of restricted sight distances at crest vertical curves on accident rates *(1,2)*. Although the model has not been validated by comparison with accident data, it provides a useful first approximation of benefits that can be expected from improvements to substandard vertical curves. The model is intuitively appealing because it accounts not only for the severity and length of the sight restriction, but also the relative hazard of that portion of the hidden roadway. According to Neuman and Glennon, the model is likely to overstate the detrimental effects of restricted sight distance. Pending more substantial validation, the model should be used with caution and the understanding that it probably yields an upper bound for accident reductions resulting from increasing sight distance where existing conditions do not meet American Association of State Highway and Transportation Officials (AASHTO) standards.

As postulated by the model, the number of accidents attributable to a sight restriction is directly related to both its severity and its length. The severity of the restriction is measured by the difference between the speed at which vehicles operate on the curve and the speed at which, according to AASHTO procedures, they could safely stop in emergencies. The length of the restriction is the distance over which the available sight distance is less than that considered adequate by AASHTO procedures for the actual highway operating speed.

For a highway segment containing an isolated vertical curve, the accident model can be expressed as

$$N = AR_h (L)(V) + AR_h (L_r)(V)(F_{ar}) \qquad (1)$$

where

- N = number of accidents on a segment of highway containing a crest curve,
- AR_h = average accident rate for the specific highway—or alternatively for the related general highway class—in accidents per million vehicle miles,
- L = length of highway segment in miles,
- V = traffic volume in millions of vehicles,
- L_r = length of restricted sight distance in miles (the distance over which sight distance is below or equal to the value specified by AASHTO), and
- F_{ar} = a hypothetical accident rate factor that varies according to the nature of the sight restriction and the nature of the hidden hazard.

In applying Equation 1, the average accident rate, AR_h, is preferably based on historical data collected for a substantial length of the highway under consideration. In the absence of such data, the use of statewide averages for the particular highway type is recommended as an alternative. For their cost-effectiveness evaluation, Neuman and Glennon used a rate of about 2.4 accidents per million vehicle miles to represent average conditions on two-lane rural highways *(1)*.

The length of sight restriction, L_r, is a complex function of the highway operating speed and curve geometrics. As an approximation, it can be estimated as

$$L_r = (a_0 + a_1 A)(1/5{,}280) \qquad (2)$$

where the a's are the constants identified in Table E-1 and A is the absolute value of the grade difference in percent. To use Table E-1, the highway design speed must first be determined. This requires computing the stopping sight distance (SSD) for existing curvature conditions as follows:

$$SSD = [7.017 \times 10^6 \, (L_{vc})/A]^{0.5} \qquad \text{for } SSD < L_{vc} \qquad (3)$$

or

$$SSD = 2{,}640(L_{vc}) + 664.5/A \qquad \text{for } SSD > L_{vc} \qquad (4)$$

TABLE E-1 Constants Used for Determining Length of Restricted Sight Distance (L_r) by Equation 2

Highway Operating Speed on Vertical Curve (mph)	Highway Design Speed (mph)							
	60	55	50	45	40	35	30	25
Values of a_0								
60	-524	-138	-25	113	202	256	305	382
55		-452	-163	11	111	172	221	301
50			-405	-65	45	115	169	248
45				-332	-76	21	82	167
40					-272	-55	15	110
35		No sight restriction				-231	-74	51
30							-193	19
25								-130
Values of a_1								
	207.3	152.6	120.9	80.2	56.6	38.6	29.4	15.3

where SSD is the stopping sight distance in feet and L_{vc} is the length of the vertical curve in miles. The design speed is then found by interpolation from Table E-2. The appropriate accident rate factor, F_{ar}, is selected from Table E-3.

Using Equation 1, the accidents attributable to a specific curve (excluding its straight approaches) can be estimated as

$$N_c = AR_h (L_{vc})V + AR_h (L_r) (V) (F_{ar}) \qquad (5)$$

where N_c is the number of accidents attributable to the curve. Again using Equation 1 and making comparisons as necessary over a common length of highway, the change in accidents expected from improving the stopping sight distance is as follows:

$$\Delta N = AR_h (V) [\Delta (L_r F_{ar})] \qquad (6)$$

TABLE E-2 AASHTO Stopping Sight Distance as a Function of Design Speed (3)

Design Speed (mph)	Stopping Sight Distance (ft)	Design Speed (mph)	Stopping Sight Distance (ft)
25	150	45	325
30	200	50	400
35	225	55	450
40	275	60	525

TABLE E-3 Accident Rate Factors (F_{ar})

Severity of Sight Restriction (mph)	Degree of Hazard in Sight-Restricted Area[a]		
	Minor	Significant	Major
0	0	0.4	1.0
5	(0.3)	(0.8)	(1.4)
10	0.5	1.1	1.8
15	1.2	2.0	2.8
20	2.0	3.0	4.0

NOTE: Numbers in parentheses were interpolated from Glennon *(2).*
[a]See Table E-4.

TABLE E-4 Classification of Degree of Hazard in Sight-Restricted Area

Relative Hazard	Geometric Condition
Minor	Tangent horizontal alignment Mild curvature (less than 3 degrees) Mild downgrade (less than 3 percent)
Significant	Low-volume intersection Intermediate curvature (3 to 6 degrees) Moderate downgrade (3 to 5 percent) Structure
Major	High-volume intersection Y-diverge on road Sharp curvature (greater than 6 degrees) Steep downgrade (greater than 5 percent) Narrow bridge Narrowed pavement

The fractional change in accidents is expressed by the quotient of Equations 5 and 6 as follows:

$$\frac{\Delta N}{N_c} = \frac{\Delta(L_r F_{ar})}{L_{vc} + L_r F_{ar}} \qquad (7)$$

where the denominator on the right represents conditions on the original curve and the numerator represents the change from original conditions.

The effect of curve improvements on accidents can best be estimated by applying the ratio of Equation 7 to the known number of accidents on the original curve. Should the historical accident record be unavailable, Equation 6 provides the next best estimate, the most reasonable estimate being made for AR_h.

Equations 6 and 7 are useful primarily for evaluating the safety benefits of incremental improvements in sight distance over practical ranges. They are not applicable when the actual operating speed is less than the design speed for either the unimproved or improved condition. In such a circumstance, further improvements to the design speed are not expected to result in substantial added benefits to safety.

REFERENCES

1. T. R. Neuman and J. C. Glennon. "Cost-Effectiveness of Improvements to Stopping Sight Distance Safety Problems." In *Transportation Research Record 923*. TRB, National Research Council, Washington, D.C., 1983, pp. 26–34.
2. J. C. Glennon. "Effect of Alinement on Highway Safety: A Synthesis of Prior Research." In TRB State-of-the-Art-Report. TRB, National Research Council, Washington, D.C. (forthcoming).
3. *A Policy on Geometric Design of Highways and Streets*. American Association of State Highway and Transportation Officials, Washington, D.C., 1984.

Appendix F
Relationship Between Accidents and Specific Roadside Features

Described in this appendix are the development and calibration of a roadside encroachment model, one of two types of models used to quantify the safety effects of the highway roadside environment. The second type, regression modeling, was used in work by Zegeer et al. *(1)* and is briefly summarized in Chapter 3 and Appendix C. Using gross measures of roadside condition—such as "percentage exposure length to objects within 6 m" *(2)* or "number of discrete objects, 0 to 5 ft from pavement edge" *(3)*—regression models are most useful for explaining the overall contribution of the roadside environment to highway safety. At the same time, they have been effectively used in analyzing the effects of specific types of roadside objects such as utility poles *(4)*. In contrast, roadside encroachment models are an attractive alternative for capturing the effects of a variety of roadside objects while permitting detailed investigation of either a specific highway site characterized by a homogenous roadside condition or an extended segment with a mixture of roadside elements. Because of these advantages, an in-depth investigation of encroachment models was undertaken to supplement the work by Zegeer et al. *(1)*.

Conceptually, roadside encroachment models capture the sequence of events culminating in a roadside accident and described as follows:

- An out-of-control vehicle leaves the travel lanes and encroaches on the roadside,
- Location of encroachment is such that the path of travel is directed toward a potentially hazardous object or slope,

- Hazardous object is sufficiently close to the travel lanes that control is not regained before encounter or collision between vehicle and object, and
- Collision is sufficiently severe enough to result in an accident.

When expressed in the nomenclature of mathematical probability, the encroachment model assumes the following form:

$$Ex(A_h) = Ex(E)\ Pr(E_h|E)\ Pr(C_h|E_h)\ Pr(A_h|C_h) \qquad (1)$$

where

$Ex(A_h)$ = expected annual number of roadside accidents involving a specific hazard (h);
$Ex(E)$ = expected annual number of encroachments on the highway segment encompassing the hazard (typically 1 mi long);
$Pr(E_h|E)$ = conditional probability that, given an encroachment, its location is such that an impact with the hazard is possible;
$Pr(C_h|E_h)$ = conditional probability that, given an encroachment in the potential impact area, a collision between vehicle and object will occur; and
$Pr(A_h|C_h)$ = conditional probability that, given a collision, its severity will be so great as to result in an accident.

The following extension is useful for examining more severe types of accidents:

$$Ex(CA_h) = Ex(A_h)\ Pr(CA_h|A_h) \qquad (2)$$

where $Ex(CA_H)$ is the expected annual number of casualty (injury or fatality) accidents involving the hazard and $Pr(CA_h|A_h)$ is the conditional probability that, given an accident, an injury or fatality will occur.

Development and calibration of roadside encroachment models were effectively delayed until the mid-1960s when field data describing the nature and rate of roadside encroachments became widely available (5, 6). Quickly these data were incorporated into an encroachment model used in this special case for evaluating the safety effects of luminaire supports (7). Later models (8-10) have continued to rely in large part on this original data base. Unfortunately, most of the original data were collected at a single, low-volume (from 2,000 to 6,000 vehicles per day) freeway site. The freeway had two 12-ft lanes in each direction and alignment was predominantly straight. Most important, only median encroachments beyond a 3-ft stabilized shoulder were investigated. In this study no evidence was found that any of the available models had been validated and it was concluded that, given the inappropriateness of the encroachment data, no available model could be recommended for use in analyzing the safety effects of roadside hazards on two-lane highways.

As a result, an independent calibration was undertaken based primarily on accident rather than encroachment data. Fortunately, an appropriate data base developed in the earlier investigation of utility pole accidents was available *(4)*. The data base was extensive, involving about 9,500 accidents over 2,500 mi of roadway in four states; the majority of the 1,500 highway segments was two-lane roads located in rural areas with a wide range in traffic volumes and, to a somewhat lesser extent, pole offsets were included from the travel lanes. The calibration and testing of this roadside encroachment model for two-lane, rural conditions is described in the remainder of this appendix.

Referring to Equation 1, the first required element of the model is $Ex(E)$, the expected annual number of encroachments per mile of highway. For encroachments on both sides of a two-lane roadway, the assumed model is

$$Ex(E) = \frac{Ex(EXC)}{2} [1 + Pr(Y \geq L-S)] \tag{3}$$

where $Ex(EXC)$ is the expected annual number of lane encroachments per mile and $Pr(Y \geq L)$ is the probability that an errant vehicle, veering to the left, will cross the adjacent lane of width, L, and encroach on the roadside. As used here, lane encroachment describes an out-of-control vehicle that travels onto an adjacent lane or shoulder, an occurrence assumed to be equally likely for left and right directions. The first of the two components of the bracketed term in Equation 3 is for movement to the right: here a lane encroachment always results in a roadside encroachment. The second is for movement to the left in which some probability exists for safe recovery before reaching the roadside. The expected number of encroachments, $Ex(EXC)$, is assumed to be related only to traffic volume as follows:

$$Ex(EXC) = a(ADT)^b \tag{4}$$

where ADT is the two-directional average daily traffic volume in vehicles per day and a and b are calibration constants. Refinement of Equation 4 to reflect influences of roadway elements such as curvature and lane width was impractical for this preliminary effort.

The next element of Equation 1 to be modeled is $Pr(E_h|E)$, the conditional probability that, given an encroachment, its location is such that an impact with the roadside hazard is possible. Letting X represent the distance in feet along the roadway within which an encroachment, if continued sufficiently far, will result in a collision with the hazard, then

$$Pr(E_h|E) = \frac{X}{10,560} \qquad (5)$$

The denominator of Equation 5 is changed to 5,280 when $Ex(E)$ represents encroachments on only a single roadside.

The impact envelope used for computing X is shown schematically in Figure F-1. The path of the errant vehicle is assumed to be straight and X is a

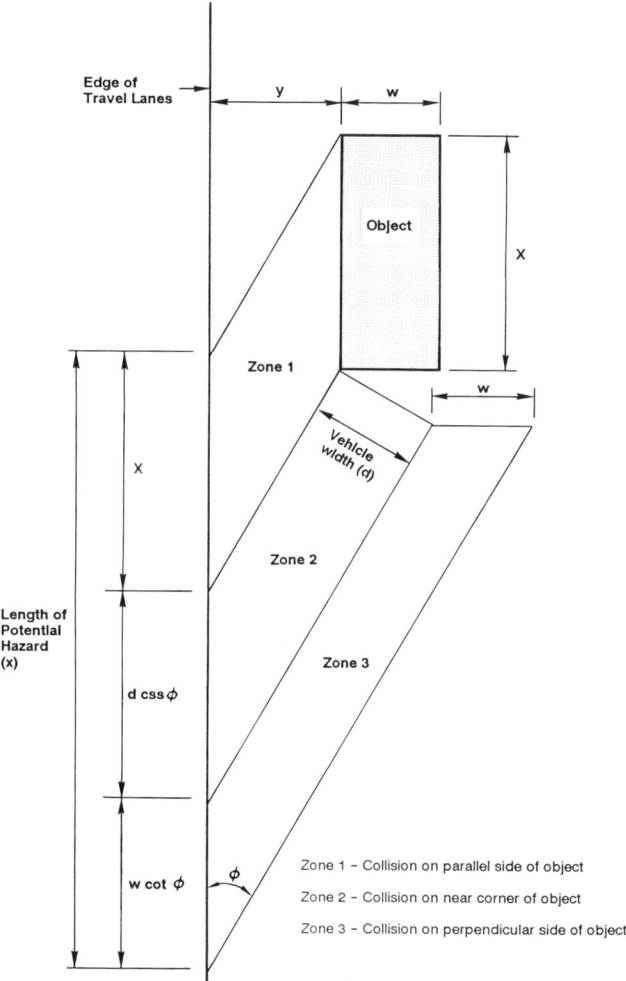

FIGURE F-1 Envelope of potential hazard based on trace of right, front corner of encroaching vehicle.

function only of the angle of departure, the dimensions of the object, and the width of the colliding vehicle. For this investigation, each utility pole was assumed to have a square cross section with 8-in. sides; the departure angle, ϕ, was taken as 6.1 degrees for right-side departures and 11.5 degrees for left-side departures *(8)*. The width of the colliding vehicle, *d*, was taken as 6 ft. Using these parameters, the projected length along the roadway of potential hazard from a single utility pole is 63.4 ft for a right-side departure and 34.0 ft for a left-side departure.

Returning to Equation 1, the next element to be modeled is $Pr(C_h|E_h)$, the conditional probability that, given an encroachment in the potential impact area, a collision between vehicle and object will occur. Again assuming that the path of travel is straight, this probability reflects the likelihood that control of the vehicle will be regained before the vehicle reaches the object. Mathematically, the expression is

$$Pr(C_h|E_h) = Pr(Y \geq y) \tag{6}$$

where $Pr(Y \geq y)$ is the probability that the vehicle—more precisely its outer front fender—will continue beyond a lateral distance of *y* from the lane boundary if its travel is not impeded by a prior collision or overturn and if control is not regained. In this investigation, the following three, one-parameter distributions of lateral travel were considered:

Linear
$$Pr(Y \geq y) = (1 - y/c) \quad \text{for } y \leq c$$
$$Pr(Y \geq y) = 0 \quad \text{for } y > c \tag{7}$$

Exponential
$$Pr(Y \geq y) = e^{c(y)} \tag{8}$$

Sinusoidal
$$Pr(Y \geq y) = \frac{1 + \cos[c(y)]}{2} \quad \text{for } y \leq 180/c \tag{9}$$

$Pr(Y \geq y) = 0$ for $y > 180/c$

where *c* is a calibration constant.

The final two necessary components of the model are $Pr(A_h|C_h)$, the conditional probability of an accident given a collision, and $Pr(CA_h|A_h)$, the conditional probability of a casualty given an accident. Although the first of these has not been extensively addressed in the literature, Zegeer and Parker *(4)* have presented usable estimates until more substantive values become available (Table F-1, accidents per collision). However, in view of the Mak and Mason finding of approximately one unreported collision with utility poles for every two reported accidents *(11)*, the Zegeer and Parker estimates *(4)* may eventually prove to be excessive.

Also presented in Table F-1 are recommendations by Glennon and Wilton for the fraction of accidents that result in a casualty, either a fatality or a nonfatal injury *(8)*. These recommendations served as a guide for extrapolating the Zegeer and Parker estimates of accidents per collision to other types of objects and are suitable for direct use with the roadside encroachment model.

Roadside encroachments in the vicinity of a utility pole are generated by near-side vehicles departing to the right and far-side vehicles departing to the left. For traffic distributed equally in both directions, the expected annual number of encroachments, $Ex(E_p)$, in the impact zone of a utility pole is

$$Ex(E_p) = a(ADT)^b \ [(0.25)(63/5,280) \\ + (0.25)(34/5,280) \ Pr(Y \geq L)] \qquad (10)$$

where the bracketed term includes both near- and far-side vehicles, 0.25 represents the fraction of total encroachments directed toward a single side of the roadway and associated with traffic moving in one direction, and L is the lane width in feet.

The expected annual number of collisions with the utility pole, $Ex(C_p)$, reflects not only the number of encroachments in the impact zone but also the lateral offset of the pole from the travel lanes. As earlier explained, increasing the offset reduces the number of collisions because of the greater chance that control can be regained before vehicles reach the pole. Examination of Figure F-1 reveals that the offset to the near-most front fender of colliding vehicles is constant for an impact at any location within Zone 1 but that it varies with the specific location of impact in Zones 2 and 3. Because "w" is small for utility poles, the offset for Zone 3 impacts can be assumed to occur at the midpoint location. Similar treatment can be given to Zone 2 by dividing it into six, 1-ft strips and using the midpoint offset of each strip. The eight segments thus considered for each direction of travel are described in Table F-2. The expected annual number of collisions with a pole can then be approximated as

$$Ex(C_p) = \frac{a(ADT)^b}{5,280}(0.25) \ [\sum_{i=1}^{8} x_i \ Pr(Y \geq y_i) + \sum_{j=1}^{8} x_j \ Pr(Y \geq y_j)] \qquad (11)$$

where the subscripts i and j refer to near-side and far-side encroachments, respectively. Assuming on average 0.9 accidents occur for each collision *(4)*, the expected annual number of collisions with the utility pole, $Ex(A_p)$, is

TABLE F-1 Most Likely Consequences of Encounters with Roadside Hazards

Type of Hazard	Accidents per Collision		Casualty Accidents per Accident (8)
	Zegeer and Parker	Extrapolated (4)	
Utility pole	0.90		0.45
Trees (> 6 in.)		0.95	0.50
Rigid signposts			
Steel (≥ 6 in.)		0.95	0.50
Timber (≥ 10 in.)		0.95	0.50
Small		0.55	0.30
Breakaway		0.20	0.20
Light or signal pole			
Rigid		0.75	0.40
Breakaway		0.20	0.20
Fixed object	0.90		
Nonclear zone	0.50		
Curb	0.10		0.35
Guardrail			
Short (< 100 ft)			
Safety end		0.35	0.35
Nonsafety end		0.45	0.45
Long (> 100 ft)			
Safety end		0.30	0.30
Nonsafety end		0.35	0.35
Fill slope			
10:1	0.05		0.15
6:1	0.20		0.15
5:1		0.25	0.25
4:1	0.30		0.35
3:1	0.60		0.45
2:1 or steeper			0.60
Cut slope			
6:1	0.05		0.15
4:1	0.20		0.15
3:1	0.30		0.25
2:1	0.60		0.35
1.5:1		0.70	0.45
1:1 or steeper		0.90	0.60
Washout ditch			0.45
Culvert (lateral or longitudinal)		0.85	0.45
Raised drop inlet		0.85	0.45
Bridge abutment or pier		1.00	0.60
Roadway over bridge			
Structure			
Open gap between parallel bridges		0.95	0.50
Bridge rail			
Smooth		0.35	0.35
Parapet-type		0.40	0.40
End		0.95	0.50
Gore abutment		0.95	0.50
Retaining wall or fence		0.65	0.35
Fireplug		0.55	0.30

TABLE F-2 Length Of and Offset To Utility Pole

Zone (Figure F-1)	Segment No.	Near-Side Encroachments		Far-Side Encroachments	
		Hazard Length (ft)	Offset (ft)	Hazard Length (ft)	Offset (ft)
1	1	0.67	y[a]	0.67	y + 12.00
2	2	9.41	y + 0.50	5.02	y + 12.49
2	3	9.41	y + 1.49	5.02	y + 13.47
2	4	9.41	y + 2.48	5.02	y + 14.45
2	5	9.41	y + 3.48	5.02	y + 15.43
2	6	9.41	y + 4.47	5.02	y + 16.41
2	7	9.41	y + 5.47	5.02	y + 17.39
3	8	6.27	y + 6.30	3.29	y + 18.21

[a] y is the perpendicular distance from the face of the pole to the near side of the travel lanes.

$$Ex(A_p) = \frac{a(ADT)^b}{23{,}467} \left[\sum_{i=1}^{8} x_i \, Pr(Y \geq y_i) + \sum_{j=1}^{8} x_j \, Pr(Y \geq y_j) \right] \quad (12)$$

The data in Table F-3 were used to evaluate the three lateral travel distribution models, Equations 7-9, and to calibrate the unknown constants, a, b, and c, for each. The actual number of accidents per pole (AA_p), computed by

TABLE F-3 Mean Utility Pole Accident Frequency (Accidents per Mile per Year) as a Function of Traffic Volume, Pole Offset, and Pole Density *(4)*

ADT	Pole Density, Pole Offset (ft)				
	3.0	7.5	15.0	22.5	26.0
Low (16 Poles per Mile)					
2,800	0.38	0.30	0.12	0.05	0.03
7,200	–	0.75	0.27	0.16	0.08
12,500	1.93	0.19	0.31	0.32	0.10
19,400	0.57	0.99	0.52	0.13	0.12
38,000	2.28	2.20	0.77	0.12	0.67
Medium (40 Poles per Mile)					
2,800	1.16	0.42	0.22	0.11	0.09
7,200	1.46	0.71	0.35	0.15	0.18
12,500	1.98	0.96	0.59	0.21	0.12
19,400	2.34	1.20	0.73	0.11	0.21
38,000	2.57	2.02	1.46	0.67	–
High (63 Poles per Mile)					
2,800	0.66	0.60	0.38	0.24	0.09
7,200	2.02	1.00	0.62	0.33	0.14
12,500	1.77	0.94	0.63	0.38	0.33
19,400	3.00	1.70	0.81	–	0.29
38,000	2.99	2.13	1.05	1.15	–

dividing each mean accident frequency in Table F-3 by the corresponding pole density, replaced the expected number, $Ex(A_p)$, in Equation 12. One of the three lateral travel distributions was then selected and values of its constant, c, treated parametrically. For each value of c, a new variable Z was computed for each of the 71 data points as follows:

$$Z = 23{,}467(AA_p)/[\sum_{i=1}^{8} x_i\, Pr(Y \geq y_i) + \sum_{j=1}^{8} x_j\, Pr(Y \geq y_j)] \qquad (13)$$

such that

$$Z = a(ADT)^b \qquad (14)$$

Following logarithmic transformations, the constants, a and b, in Equation 14 were determined using least-squares procedures.

Results of the calibration process are given in Table F-4. For each type of lateral travel distribution, the best calibration is one that (a) yields a predicted mean accident frequency that equals the actual frequency, (b) produces the largest correlation coefficient in the calibration of the transformed Equation 14, that is, between ln(Z) and ln(ADT), and (c) most accurately predicts accident frequency as measured by the correlation coefficient between the actual and predicted accident frequencies. For each of the three distributions, the first two criteria are satisfied by approximately the same calibration (Table F-4). This is considered to be the calibration of choice: at this point, the third criterion—maximum correlation between actual and predicted accident frequencies—is not unduly compromised. The three models, so calibrated, are given in Table F-5 and shown in Figures F-2 and F-3.

As measured by the correlation coefficients, the three lateral travel distributions offer approximately the same accuracy. The calculated percentages of right-side departures are 73, 61, and 57 for the best exponential, linear, and sinusoidal models, respectively. Each of these compares favorably with prior research showing that 60 to 70 percent of roadside encroachments result from right-side departures *(11-12)*. A further accuracy comparison, using nine test sites specifically reserved for that purpose by Zegeer and Parker *(4)*, revealed little significant difference among the three distributions but suggested that any of the three models could be used, at least as reliably as the best Zegeer–Parker model, to estimate utility pole accidents (Table F-6). The exponential model is recommended both for its ease of use and for the intuitive appeal of its greater sensitivity to lateral offset in regions near the travel lanes (Figures F-2 and F-3).

A primary advantage of the roadside encroachment model over the regression type module as calibrated by Zegeer and Parker *(4)* is its potential

TABLE F-4 Calibration of Roadside Encroachment Model

Lateral Distance[a] (ft)	Predicted Mean Accident Frequency[b] (Utility Pole Accidents per Mile per Year)	Correlation Coefficient	
		ln(Z) versus ln(ADT)	Actual versus Predicted Accident Frequency
Linear Distribution			
20	0.830	0.716	0.878
22	1.381	0.379	0.877
24	2.131	0.206	0.869
26	1.167	0.502	0.872
28	0.937	0.667	0.866
30[c]	0.832	0.727	0.858
35	0.734	0.712	0.831
40	0.688	0.672	0.801
50	0.654	0.621	0.748
70	0.636	0.583	0.685
Exponential Distribution			
20	0.989	0.709	0.882
22	0.923	0.733	0.881
24	0.874	0.744	0.879
26	0.836	0.747	0.876
28[c]	0.807	0.745	0.872
30	0.783	0.739	0.868
35	0.742	0.720	0.855
40	0.716	0.701	0.842
50	0.685	0.668	0.815
70	0.657	0.628	0.768
Sinusoidal Distribution			
20	1.949	0.270	0.875
22	2.763	0.181	0.868
24	1.516	0.374	0.873
26	1.133	0.549	0.871
28	0.955	0.668	0.865
30[c]	0.860	0.718	0.857
35	0.745	0.715	0.830
40	0.693	0.676	0.799
50	0.653	0.619	0.737
70	0.633	0.572	0.657

[a]Lateral distance beyond which 10 percent of errant vehicles continue as determined for each distribution by its calibration constant, c.
[b]Actual mean accident frequency is 0.8 utility pole accidents per mile per year.
[c]Recommended calibration.

TABLE F-5 Summary of Calibrated Models

Lateral Travel Distribution Model	Excursion Rate (Equation 4)	Lateral Travel Distribution (Equations 7-9)
Linear	$0.05409(ADT)^{0.5802}$	$Pr(Y \geq y) = (1-y/33.33)$
Exponential	$0.07285(ADT)^{0.5935}$	$Pr(Y \geq y) = e^{-0.08224(y)}$
Sinusoidal[a]	$0.05117(ADT)^{0.5756}$	$Pr(Y \geq y) = \dfrac{1 + \cos[4.7710(y)]}{2}$

[a] The argument of the cosine function in the sinusoidal distribution is expressed in degrees.

TABLE F-6 Comparison of Predicted and Actual Utility Pole Accidents

Test Site	Pole Offset (ft)	ADT	Pole Density (Poles per Mile)	Accident Frequency (Accidents per Mile per Year)				
				Actual	Predicted			
					Zegeer-Parker	Exponential	Linear	Sinusoidal
3	20.7	3,700	25.0	0.04	0.16	0.11	0.12	0.12
7	14.0	2,150	21.7	0.07	0.16	0.12	0.15	0.16
2	24.6	14,500	41.5	0.15	0.38	0.30	0.26	0.23
8	12.2	1,875	48.6	0.23	0.38	0.29	0.36	0.38
4	22.7	4,610	36.3	0.25	0.23	0.15	0.16	0.15
6	18.1	4,900	37.9	0.57	0.28	0.24	0.29	0.29
1	4.8	41,000	19.0	1.00	1.80	1.28	1.28	1.34
9	14.9	37,100	21.6	1.48	0.83	0.60	0.72	0.74
5	1.9	11,000	76.7	3.06	2.59	3.00	2.74	2.82
Sum of squared residuals					1.47	1.01	0.89	0.85

NOTE: Actual data and predictions using Zegeer-Parker model *(4)*.

applicability to hazards other than utility poles. For general application, the expected annual number of accidents, $Ex(A_h)$, involving any roadside hazard is given by

$$Ex(A_h) = \frac{0.07285(ADT)^{0.5935}}{21{,}120} Pr(A_h|C_h) \left[\sum_i x_i e^{-0.08224 y_i} + \sum_j x_j e^{-0.08224 y_j}\right] \quad (15)$$

where all variables are as previously defined. An analysis, similar to that used for utility poles and summarized in Figure F-1 and Table F-2, is necessary to determine the x's and y's for the particular hazard under review. The probability of an accident given a collision, $Pr(A_h|C_h)$, and, if desired, the probability of a casualty accident given an accident, $Pr(CA_h|A_h)$, can be taken from Table F-1.

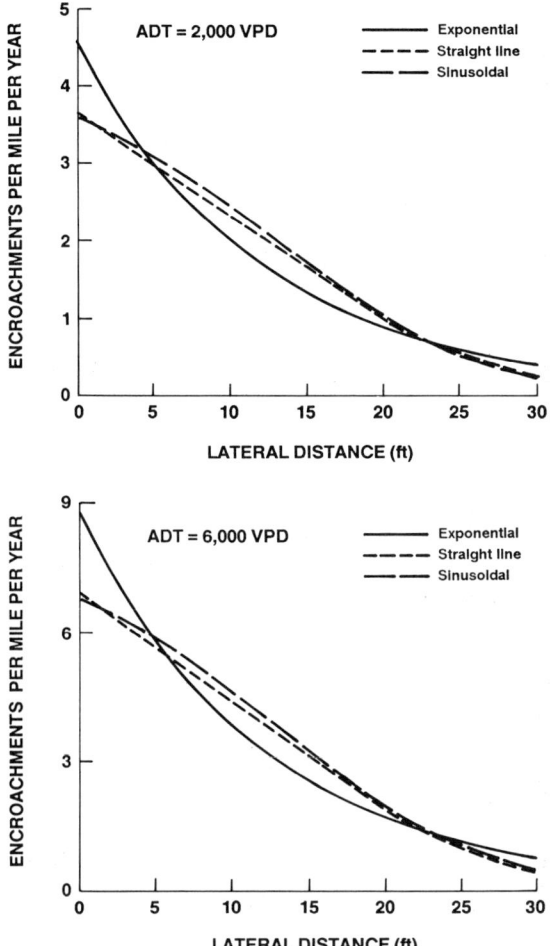

FIGURE F-2 Comparison of lateral travel distribution models on the basis of the frequency of roadside encroachments.

As a preliminary test of the general applicability of Equation 15 and Table F-1, predictions of roadside accidents for extended highway segments typical of those evaluated by Graham and Harwood *(13)* were compared with actual average conditions (Table F-7). For accidents at all levels of severity, predictions exceed actual rates by up to 160 percent, a margin believed to be only slightly magnified by the inclusion of multiple-vehicle, run-off-road accidents in the predicted quantities. For casualty accidents, the level of overprediction

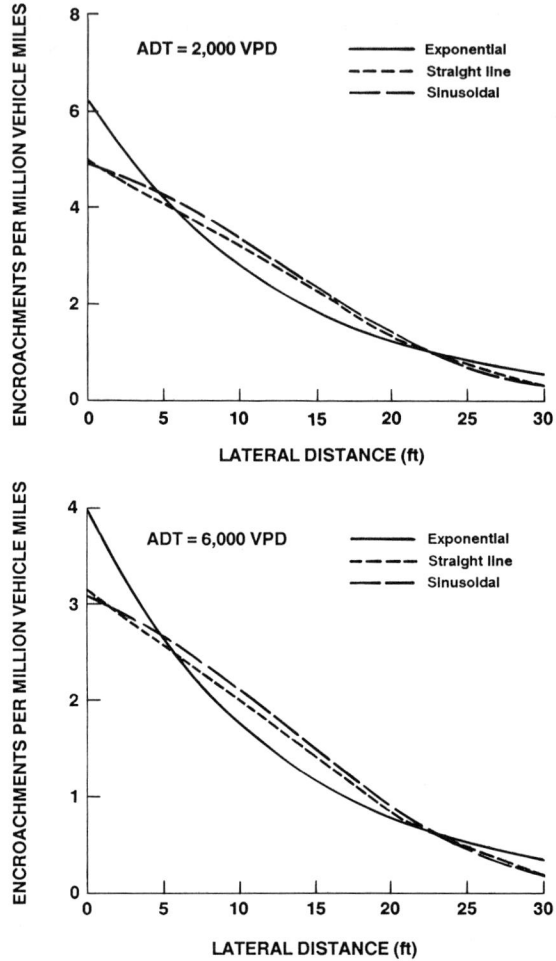

FIGURE F-3 Comparison of lateral travel distribution models on the basis of the rate of roadside encroachments.

shrank to a maximum of about 85 percent. Given the diversity of these two data sources and the numerous assumptions required in applying the encroachment model to the Graham-Harwood conditions *(13)*, agreement between predicted and actual accident rates is considered reasonably good.

One final comparison is in order, namely, between the encroachment model developed here and the models advanced earlier by other researchers. Prior models are similar only in the effect of the extent of lateral movement on

TABLE F-7 Comparison of Predicted and Actual Roadside Accidents

Clear Zone Policy[a]	All Accident Severities		Casualty Accidents	
	Predicted ROR Accident per MVM	Actual SVROR Accident per MVM	Predicted ROR Accident per MVM	Actual SVROR Accident per MVM
Nonclear zone	1.17	0.68	0.54	0.32
4:1 clear zone	0.84	0.40	0.33	0.18
6:1 clear zone	0.65	0.25	0.18	0.10

NOTES: Actual data derived from Graham and Harwood (13). ROR = run-off-road, MVM = million vehicles miles, and SVROR = single vehicle run-off-road.
[a]As defined by Graham and Harwood (13).

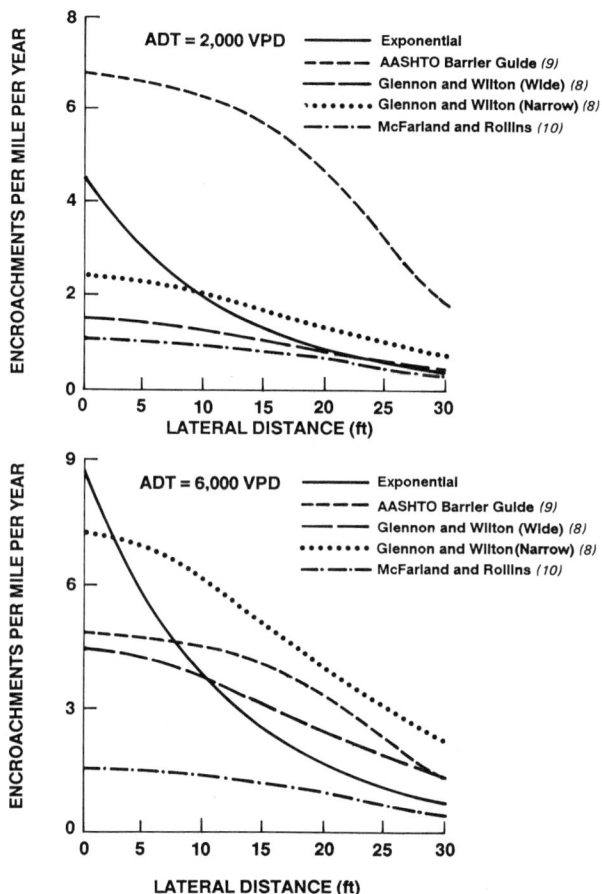

FIGURE F-4 Comparison of roadside encroachment models on the basis of the frequency of roadside encroachments.

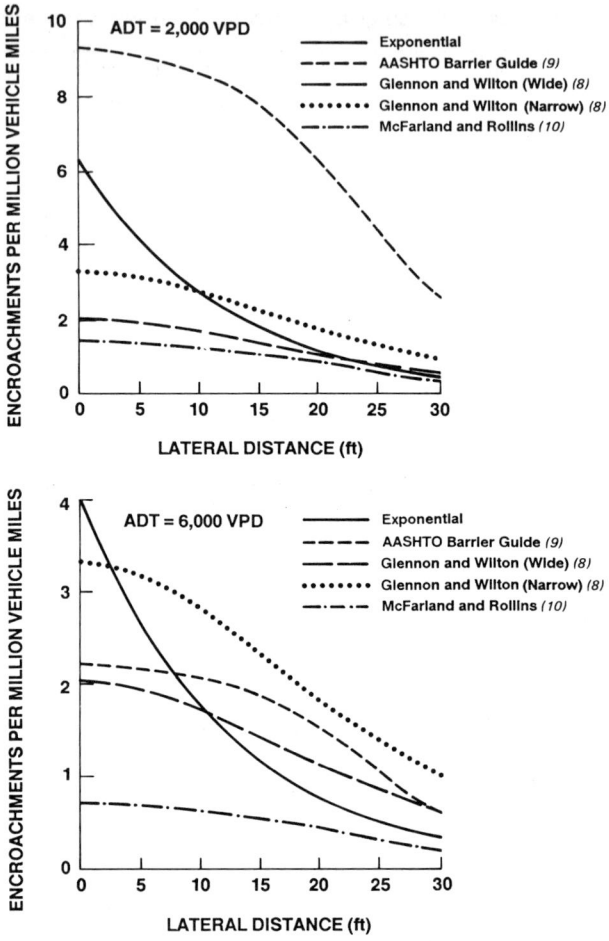

FIGURE F-5 Comparison of roadside encroachment models on the basis of the rate of roadside encroachments.

encroachment rate, a reflection of their sensitivity to the early study and analysis of freeway-median encroachments (Figures F-4 and F-5) (5-7). The exponential model developed here—as well the linear and sinusoidal models—demonstrates much greater sensitivity of encroachment rate to the extent of lateral travel. All models differ radically in their estimates of the general level of encroachments and of the influence of traffic volume on encroachment level. At a volume of 2,000 vehicles per day, the model recommended in the AASHTO *Guide for Selecting, Locating, and Designing Traffic Barriers*

(9) yields particularly extreme estimations. At all traffic volumes, the McFarland–Rollins model yields the lowest estimates of both the frequency and rate of roadside encroachments *(10)*.

Although it appears clear that much remains to be learned about roadside encroachment models, the study concluded that the model described herein was likely to be superior to other alternatives and that it could be used to provide approximate estimates of the reduction in accidents expected from roadside improvements.

REFERENCES

1. C. Zegeer, J. Hummer, D. Reinfurt, L. Herf, and W. Hunter. *Safety Effects of Cross-Section Design for Two-Lane Roads—Volumes I and II.* Report FHWA-RD-87/008 and 009. FHWA, U.S. Department of Transportation, Dec. 1986.
2. D. E. Cleveland and R. Kitamura. "Macroscopic Modeling of Two-Lane Rural Roadside Accidents." In *Transportation Research Record 681.* TRB, National Research Council, Washington, D.C., 1978, pp. 53–62.
3. P. H. Wright and K. K. Mak. "Single Vehicle Accident Relationships." *Traffic Engineering*, Vol. 46, No. 1, Jan. 1976, pp. 16–21.
4. C. V. Zegeer and M. R. Parker, Jr. *Cost-Effectiveness of Countermeasures for Utility Pole Accidents.* Goodell-Grivas, Inc., Southfield, Mich., Jan. 1983.
5. J. W. Hutchinson and T. W. Kennedy. *Medians of Divided Highways—Frequency and Nature of Vehicle Encroachments.* Bulletin 487. Engineering Experiment Station, University of Illinois, Urbana, 1966.
6. J. W. Hutchinson and T. W. Kennedy. "Safety Considerations in Median Design." In *Highway Research Record 162.* HRB, National Research Council, Washington, D.C., 1967, pp. 1–29.
7. T. C. Edwards et al. *NCHRP Report 77: Development of Design Criteria for Safer Luminaire Supports.* TRB, National Research Council, Washington, D.C., 1969.
8. J. C. Glennon and C. J. Wilton. *Effectiveness of Roadside Safety Improvements: Vol. I, A Methodology for Determining the Safety Effectiveness of Improvements on All Classes of Highways.* Report FHWA-RD-75/23. Midwest Research Institute, Kansas City, Mo., Nov. 1974.
9. *Guide for Selecting, Locating, and Designing Traffic Barriers.* American Association of State Highway and Transportation Officials, Washington, D.C., 1977.
10. W. F. McFarland and J. B. Rollins. Cost Effectiveness Techniques for Highway Safety: Resource Allocation. Final Report. Report FHWA-RD-84/011. Texas Transportation Institute, College Station, June 1985.
11. K. K. Mak and R. L. Mason. *Accident Analysis—Breakaway and Nonbreakaway Poles Including Sign and Light Standards Along Highways: Vol. II, Technical Report.* Report DOT-HS-805-605. Southwest Research Institute, San Antonio, Tex., Aug. 1980.
12. K. Perchonok et al. *Hazardous Effects of Highway Features and Roadside Objects: Vol. II, Findings.* Report FHWA-RD-78/202. Calspan Field Services, Inc., Buffalo, N.Y., Sept. 1978.
13. J. L. Graham and D. W. Harwood. NCHRP Report 247: Effectiveness of Clear Recovery Areas. TRB, National Research Council, Washington, D.C., 1982.

Appendix G
Physical and Operational Features Affecting Safety at Intersections

Although simple quantitative relationships to predict the effects of specific intersection improvements are generally not available, nonetheless, a substantial body of information exists that designers use in remedying deficiencies at intersections. This appendix contains a summary of the effects of physical and operational features on safety at intersections.

NUMBER OF LEGS

The hazard of at-grade intersections increases as the number of legs (approaches) increases. Thus, three-legged, T-type intersections are less hazardous than four-legged, cross-type intersections which, in turn, are less hazardous than five-legged intersections, and so forth. The increasing hazard results from a number of factors, including a large increase in the number of potential conflict points, an increase in the number of options underlying driver choice, difficulty in proper signing and pavement marking, including the delineation of appropriate paths of travel, and impaired surface drainage because of increased surface area within the intersection.

ANGLE OF INTERSECTION

The preferred angle between intersecting legs at an intersection is 90 degrees. At angles deviating significantly from this standard, drivers of crossing vehicles are unable to detect the presence or judge the speed of vehicles approaching on conflicting paths, and the time (and area) of potential conflict is increased. Also it becomes increasingly difficult to make turning maneuvers, in part because of larger vehicle off-tracking.

NUMBER OF THROUGH LANES

The intersection accident rate, expressed as the number of accidents per million entering vehicles, is typically greater when the approaching roadways have a larger number of lanes. Less clear is whether this phenomenon is directly the result of some aspect related to number of lanes, such as the number of potential conflict points, or whether it is indicative of indirect effects, such as larger traffic volumes. In any event, selection of the number of through lanes on intersection approaches is determined predominantly by capacity rather than by safety considerations.

SIGHT DISTANCE

In the context of at-grade intersections, sight distance refers to the view of a driver about to enter the intersection of traffic approaching on the cross road or street. As the view is enlarged by the removal of obstructions in the line of sight, the driver is better able to judge the hazard of entering the intersection, and, as a result, safety is enhanced. Although greater corner sight distance is viewed as beneficial in all circumstances, it is considerably less significant under traffic-signal control where decisions to enter the intersection are based primarily on the signal indication, and perception of threat from crossing traffic is usually a secondary consideration. Improvements to intersection sight distance impinge primarily on angle collisions.

ALIGNMENT

From a safety standpoint, intersecting roadways should be flat and straight. Curvature in either the vertical or horizontal plane that is so great as to impair sight distance will increase intersection hazard. A small gradient does not appear to be harmful and may even be advantageous in improving surface drainage. More substantial gradient becomes a liability as stopping distance on downward approaches increases, and conflict is intensified by reduced acceleration following a stop on upgrade approaches. Horizontal curvature on approaches to at-grade intersections is always harmful. Not only is it more difficult for drivers to discern the proper paths of travel but their visual focus is directed along lines tangential to these paths. Horizontal curves add further complexity to an already difficult driving environment.

AUXILIARY LANES

Auxiliary lanes reduce disruption to through traffic by accommodating special needs of turning vehicles—deceleration, acceleration, and waiting. Although auxiliary lanes are beneficial in virtually every situation, the extent of their impact is dependent on the volume of turning movements, volume of possibly conflicting movements, and approach speed. The primary impact of auxiliary lanes is on collisions between vehicles on the same approach, particularly the rear-end type. They are typically more effective in reducing hazard when accommodating left-turning vehicles than right-turning ones, and a separate left-turn phase at signalized intersections is often necessary for full benefits to be realized. One disadvantage of auxiliary lanes is increased pedestrian crossing time as a result of the added roadway width.

CHANNELIZATION

The advantages of channelization with respect to safety have been firmly established (2). Channelization can be used to delineate proper paths of travel, separate points of conflict, control angles of conflict, and provide pedestrian refuge. The establishment of left-turn lanes, a fundamental element in many channelization improvements, provides significant reductions in accident rates, particularly at unsignalized intersections (3).

Channelization is required for the provision of protected left-turn lanes, is often necessary to eliminate excessive painted areas otherwise associated with turning roadways, and is usually considered when the intersection is large. Although curbed islands provide more positive control than paved ones, hazards associated with striking raised islands suggest that painted, flush islands are preferred when approach speeds are high and fixed illumination is not provided.

TIRE-PAVEMENT FRICTION

Of all road and street locations, intersections place the greatest demands on the tire-pavement interface. The most critical conditions exist when a large number of vehicles must stop, the approach speed is great, and stopping must be accomplished quickly. The type of surfacing material, its prior wear or polishing by traffic, and the slope of the surface are important pavement attributes that affect tire-pavement friction. Grooving can be used to partly compensate for deficient pavement surfaces. Tire-pavement friction affects primarily multivehicle accidents that occur when the pavement is wet or icy.

TURNING RADII

Safety is degraded when vehicles must either encroach on adjacent lanes or slow excessively in order to execute turning maneuvers. From a geometrical standpoint, right turns are more critical than left turns, and the degree of hazard is related to both vehicle size and traffic volumes.

FIXED LIGHTING

At night, fixed lighting provides advance warning of the presence of at-grade intersections and allows the approaching driver to view objects outside the field of headlight illumination. Although the eye's delay in adjusting to the changing level of background illumination, diffusion of light on dirty or damp windshield surfaces, and the roadside hazards of lighting standards are undesirable influences, on balance the overall effect of overhead lighting is beneficial. In specific situations, such as intense pedestrian activity, installation of fixed lighting can provide substantial gains.

LANE AND SHOULDER WIDTHS

Incremental changes in lane and shoulder widths have little effect on the accident pattern at intersections. The types of accidents affected by lane and shoulder conditions on open roadways, namely, opposite-direction and run-off-road accidents, are not a significant problem at intersections. At the same time, the presence of a shoulder may well be beneficial not only because of the space it provides for collision-avoidance maneuvers but also because of sight-distance, obstacle-offset, and radius-of-turn implications.

DRIVEWAYS

Driveways in the vicinity of intersections place additional demands on the driver both in terms of the level of information that must be processed and the complexity of required decision-making. The hazard of nearby driveways, expressed primarily in their effect on multivehicle accidents, is less for entrances designed for easy and rapid use and for those located farther from the intersection. Safety is further enhanced if the site served by the driveway can properly accommodate parking and off-roadway circulation needs.

ASSIGNMENT OF RIGHT-OF-WAY

Common rules-of-the-road assign the right of intersection use to the vehicle that enters first or, in the event of two vehicles approaching simultaneously, the vehicle on the right. When sight distance is restricted or as traffic volumes increase, safety demands more positive control of traffic. Yield and stop signs and traffic signals represent progressively more definitive means for right-of-way assignment. The optimal technique from a safety standpoint depends heavily on site-specific characteristics and can best be determined by using warrants presented in the *Manual of Uniform Traffic Control Devices (1)*. Signalization, in particular, is not necessarily a safety panacea. Compared with other forms of control, signalization is often accompanied by fewer angle collisions but more rear-end collisions. The net effect of signalization depends strictly on site-specific characteristics: usually intersections having complex geometry and large traffic volumes respond best to traffic signal control. The first intersections encountered on approaches to urban areas and isolated rural intersections are difficult to safely signalize because they are not expected by approaching drivers.

APPROACH SPEED

The hazard of at-grade intersections increases as approach speed increases. Time available for driver decision-making and response is less, braking requires longer distances, and, in the event of collision, the kinetic energy dissipated is much greater. Furthermore, high approach speeds may indicate that approaching drivers do not expect the intersection and, hence, are ill-prepared to deal with the decisions it may demand. To the extent that they are capable of lowering approach speeds, traffic control devices such as advisory speed signs, flashing beacons, and rumble strips improve safety at intersections.

Higher speeds intensify the problem of dilemma zones at signalized intersections. These are zones in which motorists approaching a traffic signal that turns yellow find they are too close to stop before reaching the intersection and too far to get through the intersection before the light turns red. Special remedies that have been proposed for this problem include all-red intervals, longer yellow lights, and systems that detect the presence of a vehicle in its dilemma zone and extend the green phase *(2)*.

ON-STREET PARKING

Intersections with adjacent on-street parking are more hazardous than intersections with no parking. Vehicles parked at the roadside restrict the sideward visibility necessary for safe operation at intersections, and parking and unparking maneuvers disrupt through-traffic movements. The degree of hazard is intensified at locations of concentrated pedestrian activity.

ONE-WAY OPERATION

At least in the environment in which it is feasible, one-way operation is considerably safer than two-way operation. The advantage stems primarily from a reduction in the number of conflict points but extends to other factors such as improved signal timing, reduced headlight glare, and so forth.

MISCELLANEOUS TRAFFIC CONTROL MEASURES

Before-and-after studies demonstrate that significant safety benefits can sometimes be realized at hazardous intersections by minor changes in traffic control, including

- Improved delineation,
- No-passing signs and markings,
- Flashing beacons,
- Advance warning of intersection hazards,
- Advance directional signing,
- Prohibition of left turns,
- Enlarged signs and signal lens,
- Additional signal faces,
- Removal of roadside distractions, and
- Adjustment of signal timing.

The degree of improvement achieved by such measures is obviously dependent on the extent to which a specific deficiency can be ameliorated.

REFERENCES

1. National Advisory Committee on Uniform Traffic Control Devices. *Manual on Uniform Traffic Control Devices For Streets and Highways (as revised)*. FHWA, U.S. Department of Transportation, 1986.
2. *Synthesis of Safety Research Related to Traffic Control and Roadway Elements, Volume 1*. Report FHWA-TS-82-232. Office of Research and Development, FHWA, U.S. Department of Transportation, Dec. 1984.
3. T. R. Neuman. *NCHRP Report 279: Intersection Channelization Design Guide*. TRB, National Research Council, Washington, D.C., 1985.

Appendix H
Highway Accidents on the Federal-Aid System

The characteristics and frequency of highway accidents point not only to the role that highway design plays in safety generally, but also to the potential for accident reduction through incremental improvement to design elements. Discussed in this appendix are accident classifications, the characteristics of accidents on federal-aid systems, and accident rates.

ACCIDENT CLASSIFICATION

Highway safety is usually measured in terms of accidents or accident rates, and available data often include accident severity and accident type. Accidents generally are classified into three categories of severity: fatal, nonfatal injury, and property-damage-only accidents. Based on the number of vehicles involved, accidents are first classified into two categories by type of accident—single-vehicle and multivehicle—and then often further subdivided based on the nature of the accident. For example, single-vehicle accidents can be classified on the basis of the "first harmful event," such as hitting a fixed object or a vehicle overturn, whereas multivehicle accidents can be classified on the basis of type of vehicle interaction such as head-on collision, sideswipe, rear end, angle, and so forth.

In addition to accident severity and type, accident data are sometimes stratified in a variety of other ways—by time of day, weather and surface conditions, and driver and vehicle characteristics—depending on the purpose.

For the analysis of a specific highway feature, researchers often focus on accident types believed to be most influenced by the feature in question. For example, skid-resistance studies focus on wet-weather accidents, and roadside studies concentrate on single-vehicle, run-off-road accidents.

ACCIDENT TYPES ON FEDERAL-AID SYSTEMS

Each year the Federal Highway Administration assembles and reports nationwide estimates for fatal and nonfatal injury accidents occurring on federal-aid highway systems (Table H-1). On rural highways, nonfatal injury accidents outnumber fatal accidents by about 25 to 1. On urban streets this ratio is much higher, roughly 70 to 1, demonstrating that accidents are more severe on rural than on urban roads.

On a nationwide basis, accident data are most complete and accurate for fatal accidents; thus different federal-aid highway systems can be best compared with respect to fatal accident characteristics. Summarized in Table H-2 are the types of fatal accidents that occur on non-Interstate federal-aid systems. Differences among systems reflect variations in location (urban and rural), function (arterial versus collectors), and design. Single-vehicle accidents comprise approximately one-half to two-thirds of all fatal accidents. Of these, fixed-object accidents, particularly those involving trees and utility poles, account for about 50 to 60 percent of accidents on rural roads and about 40 percent on urban roads. These data suggest that design improvements that remove, relocate, or shield fixed obstacles would have significant potential for improved safety by reducing the severity of accidents.

Among urban multivehicle accidents, about two-thirds are rear-end or angle collisions. These accident types are particularly related to intersections and

TABLE H-1 Fatal and Nonfatal Injury Accidents on Federal-Aid Systems, 1984 *(1)*

System	Fatal Accidents	Nonfatal Injury Accidents	Ratio of Nonfatal Injury Accidents to Fatal Accidents
Rural			
Interstate	1,873	37,089	20.2
Primary	8,557	197,097	23.0
Secondary	5,573	151,047	27.1
Urban			
Interstate	1,897	97,243	51.3
Primary	3,902	287,640	73.7
Urban System	8,463	656,273	77.5

TABLE H-2 Characteristics of Fatal Accidents on Non-Interstate Federal-Aid Systems

System	Single-Vehicle Accidents				Multivehicle Accidents		
	Percent of Total	Percent Fixed Object	Percent Pedestrians	Most Common Objects	Percent of Total	Percent Head-on Collision	Percent Rear-End and Angle
Rural							
Primary	45	50	21	Trees Guardrail Utility Poles	55	48	44
Secondary	60	58	14	Trees Ditches Utility Poles	40	44	48
Urban							
Primary	50	37	46	Utility Poles Trees Guardrail	50	27	61
Arterials	55	38	46	Utility Poles Curbs Trees	45	28	62
Collectors	66	47	37	Trees Utility Poles Curbs	34	36	56

Source: 1985 Fatal Accident Reporting System Data (as summarized by L. Griffin).

turning movements, indicating that intersection improvements and access controls might also have significant potential for improved safety.

ACCIDENT RATES

Accident rates are most often calculated with respect to vehicle miles of travel (VMT), recognizing that VMT is the most readily available measure of exposure to accident occurrences. Such rates are also a convenient means of comparing the hazard associated with different highway systems (Table H-3).

From the standpoint of allocating resources for improving safety on a limited number of highway miles, however, highway agencies seek reductions in accident rates only as a step toward achieving the greatest reduction in accidents per dollar invested. Thus, the potential for enhancing safety on a given highway segment is measured in terms of accidents per year (or other unit of time), rather than accidents per VMT.

Although urban Interstates have the lowest fatal accident rates in terms of VMT, they have by far the highest rate in terms of fatal accidents per year per mile because of their high traffic volumes. Urban primaries have the next highest rate in terms of fatal accidents per year per mile (Tables H-3 and H-4).

TABLE H-3 Fatal Accidents and Fatal Accident Rates on Federal-Aid Highways [Fatal Accidents per 100 Million Vehicle Miles (MVM)] (1984) *(1)*

System	Fatal Accidents per Year	100 MVM per Year	Fatal Accidents per 100 MVM
Rural			
Interstate	1,837	1,485	1.24
Primary	8,557	2,767	3.09
Secondary	5,573	1,516	3.68
Urban			
Interstate	1,897	2,036	0.93
Primary	3,902	2,300	1.70
Urban System	8,463	3,744	2.26

TABLE H-4 Fatal Accidents and Fatal Accident Rates on Federal-Aid Highways (Fatal Accidents per Year per 100 System Miles) (1984) *(1)*

System	Fatal Accidents per Year	Highway Miles	Fatal Accidents per Year per 100 mi
Rural			
Interstate	1,837	32,676	5.62
Primary	8,557	224,868	3.81
Secondary	5,573	397,796	1.40
Urban			
Interstate	1,897	10,615	17.87
Primary	3,902	31,859	12.25
Urban System	8,463	140,492	6.02

Rural primaries and secondary roads have only about one-fifth to one-tenth as many fatal accidents per year per mile as urban Interstates. Consequently, although rural primary and secondary roads are substantially more hazardous than rural Interstates from the perspective of accidents per vehicle mile traveled, their relatively low potential for accident reduction per year limits the level of investment that can be justified for safety reasons alone.

REFERENCE

1. *Highway Safety Performance—1984, Fatal and Injury Accident Rates on Public Roads in the United States.* FHWA, U.S. Department of Transportation, Jan. 1986.

Appendix I
Initial Cost to Flatten Highway Curves

The development of models for estimating the cost of flattening highway curves is described in this appendix. Using the engineering approach, the cost models are based on estimates of hypothetical quantities of typical pay items and their related unit construction costs. Unit construction costs were based on data from the Washington State Department of Transportation for RRR projects and are considered representative of 1983 conditions. In developing "most reasonable" estimates, an attempt was made to incorporate possible economies of scale that can result in lower unit costs as construction quantities increase. However, costs are estimated as though curve reconstruction is not part of a larger RRR project. Pay items and unit construction costs are given in Table I-1, and the typical cross section defining assumed conditions is shown in Figure I-1.

Construction quantities for flattening horizontal curves and a brief explanation of the assumptions necessary for making the estimates are given in Table I-2. These quantities were used with the unit cost data in Table I-1 to estimate

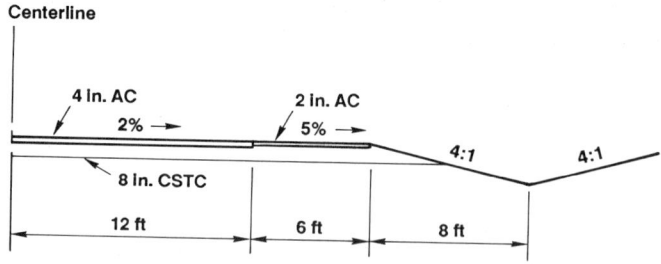

FIGURE I-1 Typical cross section used for developing cost models.

TABLE I-1 Pay Items and Unit Construction Costs

Item	Unit Cost ($)	Simplification for Vertical Curve Cost Model	Other Costs
Earthwork	Cost/CY = 10-0.00025 CY for CY ≤ 14,000 Cost/CY = 49,000/CY + 3 for CY > 14,000		
Surfacing	Cost/T = 30.5-0.00027T	$27/T	
Base (CSTC)	Cost/T = 10.7-0.00014T	$9/T	
Drainage			15 percent of earthwork, surfacing, and base costs
Clearing and grubbing	Cost/A = 2,500		
Seed, fertilize, and mulch	Cost/A = 975-11.15A	$700/A	
Guardrail	Cost/LF = 22-0.001 LF		
Pavement striping	Cost/LF = 0.06		$400 fixed cost
Wire fence	Cost/LF = 2		
Traffic control, signing	Cost/PL = 500		$10,000 fixed cost for horizontal curves; $20,000 fixed cost for vertical curves
Remove fence	Cost/LF = 1		
Remove pavement	Cost/SY = 5-0.00014 SY	$3/SY	
Remove guardrail	Cost/LF = 2		
Remove — miscellaneous		$1,000/A	
Miscellaneous			32 percent of total construction and removal costs

NOTE: A = acres, CY = cubic yards, LF = lineal feet, PL = project length in 100 ft., SY = square yards, T = tons.

total construction costs for a variety of typical curve projects having central angles of 10 to 50 degrees, original curvature of up to 48 degrees, and improved or new curvature down to 2 degrees. The following model approximates results of the detailed computations:

$$C = 18,395(I)^{0.902} (D_2)^a (D_2/D_1)^b \qquad (1)$$

where

C = construction cost in 1983 dollars,
D_1 = original curvature in degrees,
D_2 = new curvature in degrees,
I = central angle in degrees,
a = $-0.0944 - 0.405\ (I)^{0.1014}$, and
b = $-0.0758\ (D_2)^{0.648}$.

For the wide range of realistic conditions investigated, the maximum error in the approximation of Equation 1 was found to be ± 10 percent.

Construction quantities for lengthening crest vertical curves and a brief explanation of the assumptions necessary for making the estimates are given in Table I-3. Excavation quantities were estimated for a variety of typical

TABLE I-2 Construction Quantities for Flattening Horizontal Curves

Item	Quantity	Remarks
Earthwork	$CY = L_2(4 + 0.5D_1)$	Cross section in Figure I-1, balanced earthwork; typical centerline excavation of 5 ft at $D_1 = 9$ degrees; adjustment to reflect greater grading with poorer initial condition
Surfacing	$T = 0.7\ L_2$	4-in. pavement, 2-in. paved shoulder
Base	$T = 1.8\ L_2$	8-in. pavement, 10-in. shoulder plus wedge
Clearing and grubbing	$A = (0.011\ L_2$, right-of-way acres)	Area added to new inside right-of-way line but limited to 25 ft each side of new centerline
Seed, fertilize, and mulch	$A = \min/0.0007\ L_2$	15 ft each side of new alignment
Guardrail	$LF = (0.1 + 0.01D_2)\ L_2$	Minimum of 10 percent of new length with increases for larger D_2
Pavement striping	$LF = 3\ L_2$	Edge lines, centerlines, and non-passing zones
Wire fence	$LF = 2\ L_2$	Both sides
Remove fence	$LF = 2\ L_1$	Both sides
Remove surfacing	$SY = 4\ L_1$	36 ft of surfacing
Remove guardrail	$LF = (0.2 + 0.01D_1)\ L_1$	Slightly greater removal than installation
Removal—miscellaneous	$A = 0.0009\ L_1$	40 ft, shaping, cleanup, seeding, and so forth

NOTE: A = acres, CY = cubic yards, D_1 = original degree of curvature, D_2 = new degree of curvature, L_1 = feet along old alignment between points of beginning and end of new curve, L_2 = feet from beginning to ending of new curve, LF = lineal feet, SY = square yards, T = tons.

TABLE I-3 Construction Quantities for Lengthening Crest Vertical Curves

Item	Quantity	Remarks
Earthwork	Equation 2	Cross section in Figure I-1, all excavation; grade and ground elevations the same at beginning and end of both curves
Surfacing	T = 0.7 L_2	4-in pavement, 2-in. paved shoulder
Base	T = 1.8 L_2	8-in. pavement, 10-in. shoulder plus wedge
Clearing and grubbing	A = 0.0014 L_2	30 ft each side
Seed, fertilize, and mulch	A = 0.0011 L_2	25 ft each side
Guardrail		None
Pavement striping	LF = 3 L_2	Edge lines, centerlines, and nonpassing zones
Wire fence	LF = 2 L_2	Both sides
Remove fence	LF = 2 L_2	Both sides
Remove surfacing		None
Removal — miscellaneous		None

NOTE: A = acres, L_2 = feet from beginning to end of new curve, LF = lineal feet, T = tons.

improvement projects, including those having grade differences of up to 14 percent, initial curve lengths as small as 200 ft, and new curve lengths as large as 1,800 ft. The estimates ranged from a low of 452 yd³ to a high of 98,923 yd³. The following model was found to reproduce the computations with an error no greater than ± 10 percent:

$$CY = a\,(L_2^b - L_1^b) \qquad (2)$$

where

CY = cubic yards of excavation,
L_1 = original vertical curve length in feet,
L_2 = new or improved vertical curve length in feet,
G = absolute value of the percent difference in grades,
a = 0.10392 $(G)^{-1.0615}$, and
b = 1.4385 $(G)^{0.16144}$.

The simplified construction cost model for crest curve projects is summarized as

$$C = 1.52\,(CY)\,(UC) + 73.62\,L_2 + 26{,}928 \qquad (3)$$

where C is the construction cost in 1983 dollars and UC is the unit cost of excavation as given in Table I-1.

Acquisition of additional right-of-way is an anticipated consequence of horizontal curve projects and, depending on initial right-of-way width, may be required on vertical curve projects as well. However, the extreme variation in right-of-way costs makes them impossible to model with any reliability; thus, they have been omitted from Equations 1 and 3.

Appendix J

Relationship Between Cost per Accident Eliminated and Benefit-Cost Ratio Approaches

Illustrated in this appendix is the relationship between cost-effectiveness analyses for a hypothetical horizontal curve improvement using cost per accident eliminated and the benefit-cost ratio approach.

Assumed conditions for the example are

- 1,000 ADT,
- 20-degree central angle,
- 35 mph design speed before improvement,
- 55 mph design speed after improvement,
- 0.20 combined side friction and superelevation factor,
- 7 percent discount rate,
- 30-year project life, and
- $30,000 value imputed to each accident eliminated.

The imputed value for each accident eliminated is required only for the benefit-cost approach. As noted in Chapter 5, cost per accident eliminated was selected as the principal measure of cost-effectiveness (instead of the benefit-cost ratio) to avoid the arbitrary imputation of dollar values to accidents eliminated.

To determine cost per accident eliminated

- *Estimate accidents eliminated*—0.10/year, based on assumed conditions before and after improvement and the accident relationship for horizontal curves presented in Appendix E;
- *Estimate added cost*—$111,000, based on assumed conditions and the cost relationship presented in Appendix I;
- *Annualize the added cost*—$8,950/year, based on the assumed discount rate and project life; and
- *Calculate cost per accident eliminated*—$89,500, calculated as 8,950/0.10.

The preceding calculation does not account for the highway user travel time and operating cost savings associated with flattening horizontal curves. These savings can be taken into account within the cost per accident eliminated framework by subtracting them from the cost required to implement the improvement:

- *Estimate user savings*—$6,150/year, based on the methodology presented in AASHTO's *Manual on User Benefit Analysis of Highway and Bus-Transit Improvements (1)*;
- *Calculate net (implementation minus user) cost*—$2,800/year, calculated as 8,950 − 6,150; and
- *Calculate net cost per accident eliminated*—$28,000, calculated as 2,800/0.10.

For the same horizontal curve, the benefit-cost ratio is calculated as follows:

- *Estimate accidents eliminated*—0.10/year, as previously;
- *Express safety benefits in dollar terms*—$3,000/year, based on the imputed value of $30,000 per accident eliminated;
- *Estimate highway user travel time and operating cost savings*—$6,150, as previously;
- *Calculate total benefits*—$9,150/year, calculated as 3,000 + 6,150;
- *Calculate the annualized added cost*—$8,950, as previously; and
- *Calculate benefit-cost ratio*—1.02, calculated as 9,150/8,950.

As shown by this example, when net cost per accident eliminated is close to the value imputed to each accident eliminated—$28,000 versus $30,000 in this case—the benefit-cost ratio is close to 1.00.

REFERENCE

1. *Manual on User Benefit Analysis of Highway and Bus-Transit Improvements.* American Association of State Highway and Transportation Officials, Washington, D.C., 1977.

Appendix K

Effects of Lane and Shoulder Widths on Travel Time

Narrow lanes and shoulders on two-lane roads cause motorists to drive closer to vehicles in the opposing lane. They must compensate for driving closer to opposing traffic by slowing down and allowing larger headways between vehicles in the same lane. Thus, drivers experience more delay and lower speeds on roads with narrow lanes and shoulders than on roads with wider lanes and shoulders.

A procedure for estimating the effects of lane and shoulder width improvements on travel time is presented in this appendix. This procedure is based on methodology presented in the *Highway Capacity Manual (1)* and accounts for the combined effects of lane and shoulder width and traffic volumes. These combined effects can be important because the effects of narrow lanes and shoulders on speed are exacerbated when traffic volumes are greater.

PROCEDURE

For two-lane rural highways, the effects of lane and shoulder width and average daily traffic (ADT) on travel time can be estimated as

$$VHT/VMT = (1/58) \exp [0.01 \, ADT/(SFD \, f_w)] \quad (1)$$

In this equation

- *VHT/VMT* is hours per vehicle mile—the inverse of speed;
- exp is the exponential operator;
- *ADT* is annual average daily traffic;
- *SFD* is the maximum Level of Service D flow rate two-way (in vehicles per hour) for the facility under consideration, assuming 12-ft lanes and 6-ft shoulders. The value of *SFD* depends on terrain, percent no-passing, vehicle mix, and directional split of traffic. If information on these various factors is not available, a default value of 819 can be used for *SFD*; and
- f_w is a width adjustment factor used to adjust capacity downward if lane widths are less than 12 ft or shoulder widths are less than 6 ft. Values of f as a function of lane and shoulder width are given in Table K-1. These values can be approximated with the equation

$$f_w = \min [1, 0.085 \ (LW + SW/2) - 0.275] \qquad (2)$$

where *LW* is one-way lane width (in feet) and SW is one-way shoulder width (in feet).

The assumptions used in developing the default value of 819 for *SFD* in the preceding equation are as follows:

- Rolling terrain;
- No-passing zones, 40 percent;
- Directional distribution for traffic is 60–40;
- Vehicle mix is 14 percent trucks, 4 percent recreational vehicles (RVs), and 1 percent buses;
- Design speed is 60 mph or greater; and
- Speed limit is 55 mph.

Table K-2 provides alternative values of *SFD*, which can be used if specific information on terrain, percent no-passing, directional distribution, or vehicle mix is available for the highway segment under consideration.

TABLE K-1 Lane and Shoulder Width Adjustment Factors

Usable Shoulder Width (ft)	12-ft Lanes	11-ft Lanes	10-ft Lanes	9-ft Lanes
6	1.00	0.93	0.84	0.70
4	0.92	0.85	0.77	0.65
2	0.81	0.75	0.68	0.57
0	0.70	0.65	0.58	0.49

SOURCE: *Highway Capacity Manual,* Table 8-5 *(1).*

The "free flow" speed (i.e., the speed at low traffic volumes) of 58 mph is based on an assumed design speed of 60 mph or greater and a speed limit of 55 mph. For lower design speeds, the *Highway Capacity Manual (1)* recommends that speeds be reduced by 4 mph for each 10-mph reduction in design speed below 60 mph.

Application of this procedure to hypothetical RRR projects indicates that travel time savings can be an important consideration in making decisions about lane and shoulder widening, particularly on highways with more than 2,000 ADT. However, as noted in Chapter 5, these savings will be partly offset by higher motor vehicle operating costs at higher speeds. Also, higher speeds can lead to increases in the severity of accidents.

DERIVATION

Equation 1 is based on methodology for estimating level of service and speed on two-lane rural highways in Chapter 8 of the *Highway Capacity Manual (1)*. The 0.01 and 1/58 constants in the equations were derived using the Worksheet for General Terrain Segments, *Highway Capacity Manual (1*, pp.8–31), together with the default assumptions presented here.

The worksheet provides a basis for determining average speed as a function of hourly flow rate for a given highway segment. Specifically, it provides five points on a speed-volume curve. Speeds at other traffic volumes can then be determined by interpolation.

Because the manual provides speeds as a function of hourly flow rate, it is necessary to assume a temporal distribution of traffic, in order to calculate travel time as a function of ADT. For this purpose, a representative rural weekday traffic distribution was used (2, p.167).

The relationship between travel time and ADT was investigated for three cases: 10-ft lanes with 2-ft shoulders, 11-ft lanes with 4-ft shoulders, and 12-ft lanes with 6-ft shoulders. The values of f_w and the Level of Service D maximum flow rates for the three cases are

Lane Width	Shoulder Width	f_w	Level of Service D Maximum Flow
10	2	0.68	557
11	4	0.85	696
12	6	1.00	819

For each of the three cases, travel time (per mile) was calculated for ADT ranging from 500 to 7,900 (in steps of 200), and regression analysis was applied to estimate coefficients for the equation

$$\text{Ln } (VHT/VMT) = a + b \, (ADT) \tag{3}$$

The three values of "b" obtained were

- 1.80×10^{-5} for 10-ft lanes with 2-ft shoulders,
- 1.50×10^{-5} for 11-ft lanes with 4-ft shoulders, and
- 1.28×10^{-5} for 12-ft lanes with 6-ft shoulders.

In all three cases, the value of "b" can be approximated as 0.01 divided by the Level of Service D maximum flow rate.

The value of "a" was –4.06 in all three cases. This is consistent with a free flow speed of 58 mph; that is, when ADT is zero, the inverse of speed is calculated as exp (–4.06) or 1/58.

The alternative values of SFD given in Table K-2 were calculated from the *Highway Capacity Manual* (Tables 8-1, 8-4, and 8-6). Table K-2 provides

TABLE K-2 Level of Service D Maximum Flow Rates

	Percent No-Passing Zone					
	0	20	40	60	80	100
7 Percent Trucks, 2 Percent RVs, and 0.5 Percent Buses						
Level	1,553	1,504	1,455	1,431	1,407	1,383
Rolling	1,222	1,123	1,025	946	907	847
Mountainous	811	699	629	560	518	462
14 Percent Trucks, 4 Percent RVs, and 1 Percent Buses						
Level	1,440	1,395	1,350	1,327	1,305	1,282
Rolling	977	898	819	756	725	677
Mountainous	553	476	429	381	352	314
21 Percent Trucks, 6 Percent RVs, and 1.5 Percent Buses						
Level	1,342	1,300	1,258	1,237	1,216	1,195
Rolling	813	748	682	630	603	564
Mountainous	419	361	325	289	267	238

NOTES: The maximum flow rates are for 12-ft lanes with 6-ft shoulders. The width adjustment factors given in Table K-1 are applied to these flow rates if lanes are less than 12 ft or shoulders are less than 6 ft. The maximum flow rates were calculated for an assumed 60-40 directional distribution during individual hours of the day. For a 50-50 directional distribution, the flow rates are multiplied by 1.057. For a 70-30 directional distribution, the flow rates are multiplied by 0.943.

values of SFD for three different vehicle mix distributions, corresponding to high, medium, and low shares for the three types of heavy vehicles. The information given in Table 8-6 of the manual can be used to develop more precise estimates of SFD for a given vehicle mix.

REFERENCES

1. *Special Report 209: Highway Capacity Manual.* TRB, National Research Council, Washington, D.C., 1985.
2. *Allocation of Life Cycle Highway Pavement Costs.* Report FHWA-RD-83/080. FHWA, U.S. Department of Transportation, 1983.

Appendix L

Alternative Lane and Shoulder Width Standards Used in System-Level Analyses

Described in this appendix are the four sets of lane and shoulder width standards examined in the national system-level analyses presented in Chapter 5. These standards apply to two-lane rural highways only.

The American Association of State Highway and Transportation Officials' (AASHTO) new construction standards (Table L-1) are the most stringent standards analyzed in this study. For highways with design speeds of 60 mph and above, minimum lane widths are 12 ft for arterials and 11 or 12 ft for collectors, depending on traffic volumes. Under these standards shoulder widths for high-volume collectors and arterials are set at 8 and 10 ft, respectively.

AASHTO resurfacing, restoration, and rehabilitation (RRR) standards (Table L-2) are the most lenient examined; they never require more than 11-ft lane and 2-ft shoulder widths and are the only standards that do not account for traffic volume.

The 1978 FHWA proposed standards (Table L-3) usually fall between AASHTO new construction and AASHTO RRR standards in terms of stringency. Lane widths vary depending on traffic volumes, percent trucks, and design speed. Lane widths can be as low as 9-ft on low-volume, low speed "minor" roads and increase to 12 ft for highways with greater traffic volumes and higher speeds. Shoulder widths range from 2 to 4 ft.

The modified 1978 FHWA standards (Table L-4) were developed to improve the original 1978 FHWA proposed standards. The committee found

TABLE L-1 AASHTO New Construction Standards for Lane and Shoulder Width *(1)*

	Width of Road Feature (ft) by Projected Design Traffic Volume				
Design Speed	ADT Under 400	ADT 400 and Over	DHV 100–200	DHV 200–400	DHV Over 400
Arterials: Lanes					
50	11	12	12	12	12
60	12	12	12	12	12
70	12	12	12	12	12
Arterials: Shoulders					
All speeds	4	6	6	8	10
Collectors: Lanes					
20	10	10	10	11	12
30	10	10	10	11	12
40	10	11	11	11	12
50	10	11	11	12	12
60	11	11	11	12	12
70	11	11	11	12	12
Collectors: Shoulders (Graded)					
All speeds	2	4	6	8	8

NOTES: Usable shoulders should be paved; graded shoulders need not be paved; 4 ft is minimum shoulder width if a roadside barrier is used; and DHV is design hour volume, usually the traffic volume projected for the 30th highest hour of the design year. DHV is generally 10 to 15 percent of ADT.

TABLE L-2 AASHTO RRR Standards for Lane and Shoulder Widths *(2)*

		Minimum Widths (ft)	
Average Running Speeds (mph)	Percent Trucks	Lane	Shoulder
40 or less	All	10	2
Over 40	15 or less	10	2
	Over 15	11	2

NOTE: The standards specify a range of widths for all cases. Only the minimums are reported here.

that the concept of stratification by average daily traffic (ADT), truck composition, and design speed was sound but that minor changes could reduce cost per accident eliminated; this was verified by subsequent analyses. The modified standards used in the national system-level analyses differ from the original 1978 FHWA standards:

1. All ranges were shifted,
2. Shoulder widths for high-volume highways were increased by 2 ft,

3. Shoulder widths for highways in mountainous terrain were reduced by 1 ft,

4. Average running speed was substituted for design speed because the concept of design speed is difficult to apply to an existing highway, and

5. The concept of minor roads was eliminated because it was not clearly defined in the original 1978 FHWA proposal.

The first three changes were made to improve cost-effectiveness; the last two were made to clarify the standards for users.

TABLE L-3 1978 FHWA RRR Proposed Standards for Lane and Shoulder Widths (3)

Current Traffic (ADT)	Design Speed (mph)	Width (ft)			
		10 Percent or More Trucks		Less Than 10 Percent Trucks	
		Lanes	Shoulders	Lanes	Shoulders
1–400	50 or less	10	2	9[a]	2
	Over 50	10	2	10	2
401–4,000	50 or less	11	2	10	2
	Over 50	12	3	11	3
Over 4,000	All	12	4	11	4

NOTES: The 1978 FHWA proposed standards were actually defined in terms of lane width and total roadway (lane plus shoulder) width. The standards are shown here in terms of lane and shoulder width so that they can be more easily compared with AASHTO standards. In the actual standard, 11-ft lanes and a 1-ft shoulder are permitted where 10-ft lanes and 2-ft shoulders are specified in the table.
[a] "Minor roads" only; otherwise 10-ft lanes.

TABLE L-4 Modified 1978 FHWA RRR Standards for Lane and Shoulder Widths

Design Year Volume (ADT)	Average Running Speed (mph)	Width (ft)			
		10 Percent or More Trucks		Less Than 10 Percent Trucks	
		Lanes	Shoulders	Lanes	Shoulders
1–750	Under 50	10	2	9	2
	50 or more	10	2	10	2
751–2,000	Under 50	11	2	10	2
	50 or more	12	3	11	3
Over 2,000	All	12	6	11	6

NOTES: Shoulders may be 1 ft less for highways in mountainous terrain. The standards were actually defined in terms of lane width and total roadway (lane plus shoulder) width. They are given here in terms of lane and shoulder width for easier comparison with AASHTO standards. For the purpose of simplicity, weighted average design speed was used for average running speed in the analysis.

REFERENCES

1. *A Policy on Geometric Design of Highways and Streets*. American Association of State Highway and Transportation Officials, Washington, D.C., 1984.
2. *Geometric Design Guide for Resurfacing, Restoration and Rehabilitation (RRR) of Highways and Streets*. American Association of State Highway and Transportation Officials, Washington, D.C., 1977.
3. Design Standards for Highways. *Federal Register*, Vol. 43, No. 164, Aug. 23, 1978.

Study Committee Biographical Information

PETER G. KOLTNOW, *Co-Chairman*, is a consulting engineer and Counselor to the President, American Trucking Associations. He received his bachelor's degree from Antioch College and his master's degree from the University of California. Mr. Koltnow has worked in transportation and traffic safety for both the public and private sectors during his career. He served as chairman of the TRB Executive Committee in 1979, and has served on the Executive Council of the Institute for Transportation, American Public Works Association, and the Executive Committee, National Committee on Uniform Traffic Laws and Ordinances. He was President of the Highway Users Federation for Safety and Mobility from 1974 to 1984. A registered professional engineer in California and Ohio, Mr. Koltnow is a member of the American Society of Civil Engineers and a Fellow of the Institute of Transportation Engineers.

HERBERT H. RICHARDSON, *Co-Chairman*, is Deputy Chancellor for Engineering, and Director, Texas Engineering Experiment Station, Texas A&M University System, where he also holds the positions of Dean of Engineering and Distinguished Professor of Engineering. Dr. Richardson earned bachelor's and master's degrees from the Massachusetts Institute of Technology, where he also completed his doctor of science degree in mechanical engineering, graduating in 1950. Before joining Texas A&M University, Dr. Richardson was Professor and Associate Dean of Engineering at MIT, becoming Professor Emeritus in Mechanical Engineering in 1984. During this period he also served with the U.S. Army Ordnance Corps at the Ballistics Research Laboratory, Aberdeen Proving Grounds in Maryland. Before joining the faculty of MIT, he was Chief Scientist, Office of the Secretary of the U.S. Department of Transportation. To this background, he adds varied experience in consulting, listing the Engineering Societies Commission on Energy, International Harvester, Inc., Foster-Miller, Inc., and the Caterpillar Tractor Company among his many clients. Currently, Dr. Richardson also serves on numerous committees and panels, such as the Council, National Academy of Engineering; the Committee on Engineering Education, National Academy of Engineering; the Governing Board of the National Research Council; and the Executive Committee of the Transportation Research Board. He is also a member of the National Science Foundation's Engineering Advisory Committee. Dr. Richardson is active in the work of the American Society of Mechanical Engineers and in that of other scientific societies. He was a cowinner of the Moody Award in 1970, and a recipient of the Pi Tau Sigma Gold Medal for "Outstanding Mechanical Engineer Ten Years After Graduation" in 1963. He was also awarded the ASME Centennial Medallion in 1980 and the Rufus Oldenberg Medal, ASME, in Dynamical Systems and Control in 1984. Dr. Richardson is the author of many publications, contributing a number of books and innumerable major reports and magazine articles to his field.

ROY W. ANDERSON, a registered professional engineer, is currently President of TranSafety, Inc. He received his bachelor's degree from Texas Tech University and his master's degree from the University of California, Berkeley. Mr. Anderson has served as Director of the Office of Safety Studies, National Transportation Safety Board, and has worked as a civil engineer for the California Department of Transportation. He was a member of the National Cooperative Highway Research Program Panel on Evaluation of Traffic Controls for Street and Highway Work Zones from 1978 to 1980. A former member of the FHWA Safety Review Task Force, Mr. Anderson is the recipient of the California Society of Professional Engineers' award for outstanding performance and a special service award from the National Transportation Safety Board. He is a former state director of the National Society of Professional Engineers, a member of the American Association for Automotive Medicine, and a member of the American Society of Civil Engineers. He also serves on the Safety Coordinating Committee of the Institute of Transportation Engineers.

LEONARD EVANS is a physicist and Principal Research Scientist in the Operating Sciences Department of the General Motors Research Laboratories. A graduate of Queens University, Belfast, and Oxford University, Dr. Evans' more that 70 technical publications cover such diverse subjects as physics, mathematics, traffic engineering, transportation energy, human factors, trauma analysis, and traffic safety. His main professional interests focus on traffic safety research. He is a member of the Society of Automotive Engineers (SAE) and is past chairman of the SAE Human Factors Committee. Dr. Evans is also a member of the Human Factors Society, having served as chairman of its Southeastern Michigan Chapter. He is a member of the American Association for Automotive Medicine, Sigma Xi, the American Association for the Advancement of Science, the Society for Risk Analysis, and the Editorial Advisory Boards of *Accident Analysis and Prevention* and *Human Factors*.

JOHN C. GLENNON is a transportation consulting engineer. He is a graduate of the University of Illinois with a bachelor's and a master's degree, and the University of Kansas with a DE degree. Dr. Glennon joined the California Division of Highways where he worked as an Assistant Highway Engineer and later as Assistant State Transit Planning Coordinator before joining the Texas Transportation Institute as a Research Engineer. Following his tenure at the Texas Transportation Institute, Dr. Glennon joined the Midwest Research Institute (MRI) where he served first as Manager of the Traffic Safety Center and later as Manager of the Design and Operation Program. He is a member of the TRB Task Force on Tort Liability and a former member of the TRB Committee on Vehicle Characteristics. He was a member and subcommittee chairman of the TRB Committee on Operational Effects of Geometrics from 1972 to 1980, and is currently Chairman of the TRB Committee on Geometric Design. In addition, he has chaired the Special Advisory Panel to Review FHWA Research Programs on Geometric Design, the ITE Committee on Criteria for Installation of Concrete Barrier Wall Versus Guardrail, and the ITE Committee to Review Standards for Urban Arterial Streets. Dr. Glennon has also served on several FHWA advisory panels dealing with such topics as size and weight of heavy vehicles and highway barrier need indices. He received the TRB D. Grant Mickle award (with D. Harwood) for an outstanding paper on highway traffic and maintenance and the ITE Missouri Valley Section President's Award for his outstanding contribution to traffic engineering. Dr. Glennon is a member of the National Society of Professional Engineers, National Association of Forensic Engineers, American Academy of Forensic Sciences, and the Institute of Transportation Engineers.

EZRA HAUER is an engineer and Professor, Department of Civil Engineering, University of Toronto. He received his bachelor's and master's degrees from Technion University in Israel and his Ph.D. from the University of California. He has served as a special consultant to the Midwest Research Institute; the World Bank; the Department of Transport, Road, and Motor Vehicle Traffic Safety; and DeLeuw, Cather and Company. Dr. Hauer is a member of the Association of Professional Engineers of Ontario, Institute of Transportation Engineers, American Association for Automotive Medicine, Operations Research Society of America, Canadian Operations Research Society, and the Israeli Association of Transportation Research.

W. RONALD HUDSON is a Professor of Civil Engineering and Dewitt C. Greer Centennial Professor of Transportation at the University of Texas, Austin. He received his bachelor's degree from Texas A&M University and his master's degree and Ph.D. from the University of Texas. Dr. Hudson served as Research Engineer for the Highway Research Board AASHO Road Test from 1958 to 1961. He also served as Research Engineer for the Center for Highway Research and as Design Research Engineer for the Texas Highway Department. He was an Assistant Project Engineer for the National Cooperative Highway Research Program from 1963 to 1964. Since 1979 he has been Chairman of TRB's Pavement Management Section. He is a member of the TRB Committee on Monitoring, Evaluation, and Data Storage and the Executive Committee, Highway Division, American Society of Civil Engineers (ASCE). Dr. Hudson received the Highway Research Board's Outstanding Paper Award in 1964 and the ASCE Texas Section's Outstanding Paper Award in 1965. He is a member of the American Association for the Advancement of Science, National Society of Professional Engineers, American Concrete Institute, and the New York Academy of Sciences. Dr. Hudson has conducted research on pavement management and performance, improved and stabilized materials, and application of economics and statistical methods to engineering problems. He is the author of *Pavement Management Systems* (McGraw-Hill, 1978).

JACK T. KASSEL, an engineer, recently retired from the position of Project Development Division Chief for the California Department of Transportation. Mr. Kassel received his bachelor's degree from the University of California, Los Angeles, and has had a long, distinguished career at the California Department of Transportation, having served as Project Design Engineer, Los

Angeles District; District Traffic Engineer; Assistant Traffic Engineer, Division of Highways; Computer Systems Engineer; Chief, Office of Local Assistance, Caltrans; Chief, Office of State Planning; Chief, Division of Equipment; and Chief, Division of Value Engineering. Mr. Kassel is a member of the Institute of Transportation Engineers.

JAMES L. MARTIN, an engineer and administrator, recently retired as Public Works Director for the city of Fresno, California. He holds a bachelor's degree in civil engineering from George Washington University. He began his career as a civil engineer for the U.S. Bureau of Reclamation, and has served as Bridge Engineer for the state of California; Civil Engineer, Richmond, California; Assistant City Engineer, San Leandro, California; and Public Works Director, Berkeley California. He currently serves on the Transportation Advisory Committee, California Department of Transportation, and has served on the boards of directors of the League of California Cities; the Municipal and Airports Division, American Road and Transport Builders Association; the Fresno Metropolitan Flood Control District; and the Fresno County Water Advisory Committee. Mr. Martin was a member of the board of directors of the American Public Works Association from 1976 to 1985 and president from 1983 to 1985. He is an American Society of Civil Engineers Fellow.

BROOKS O. NICHOLS is Chief Engineer of Design for the Arkansas State Highway and Transportation Department where he served as Engineer of Highway Design, Roadway Design Section Head, Assistant Roadway Engineer, and Roadway Design Engineer before assuming his current position. Mr. Nichols is a graduate of the University of Arkansas with a bachelor's degree in civil engineering. He is a member and former Chairman of the AASHTO Task Force on Geometric Design and a member of the AASHTO Subcommittee on Design and the AASHTO Joint Task Force on Pavements. Mr. Nichols is also Chairman of the Arkansas Highway and Transportation Department Research Council and a member of the University of Arkansas Academy of Civil Engineering. In 1981 he received the AASHTO Region II Design Award.

BRIAN O'NEILL is President of the Insurance Institute for Highway Safety (IIHS) and its associated organization, the Highway Loss Data Institute (HLDI), two independent organizations dedicated to reducing the losses—deaths, injuries, and property damage—resulting from motor vehicle crashes. He received his bachelor's degree in mathematics from the Bath University of Technology, England. Before becoming president of IIHS and HLDI, Mr. O'Neill served as vice president for research, senior vice president, and executive vice president of IIHS and senior vice president of HLDI. He was responsible for the research programs of both organizations, and over the years he has been personally involved in research covering virtually all aspects of highway loss reduction, including vehicle and highway design, emergency medical care, the effectiveness of traffic laws, and driver behavior. He is the author of numerous scientific papers and coauthor of the Injury Fact Book. Mr. O'Neill is Chairman of the National Safety Council's Committee on Alcohol and Other Drugs. He has served on the Advisory Committee for the U.S. Department of Transportation's National Accident Sampling System and the National Academy of Sciences' Committees on Trauma Research.

ROBERT H. RAYMOND, JR., is presently Assistant Chief Counsel in charge of the General Law Division, Commonwealth of Pennsylvania Department of Transportation. He received his AB degree from Bucknell University and his JD from Dickinson School of Law. Mr. Raymond has held the following positions in his 23 years with the Pennsylvania Department of Transportation: Chief, Environmental Section, General Law Division; Assistant Counsel, Regional Attorney, Land Acquisition Division; and Trial Attorney, Eminent Domain and Right-of-Way Division. He was a partner in the law firm of Ziegler and Raymond from 1966 to 1967. He has been admitted to practice before the U.S. Supreme Court; U.S. Court of Appeals for the Third Circuit; U.S. Court of Claims; U.S. District Court, Middle District of Pennsylvania; U.S. District Court, Eastern District of Pennsylvania; Supreme Court of Pennsylvania; Superior Court of Pennsylvania; and the Commonwealth Court of Pennsylvania.

JOHN H. SHAFER is currently an executive with the New York State Thruway Authority. He was formerly Assistant Commissioner and Chief Engineer for the New York State Department of Transportation. He received his bachelor's degree in civil engineering from the University of Detroit. He has held field and office positions in the New York Department of Public Works (Rochester) and has served as Director of the Project Development Bureau and the Safety and Traffic Division for the New York State Department of Transportation. Mr. Shafer represented the New York State Department of Transportation Commissioner on the Governor's Traffic Safety Committee and is a cofounder of the Upstate Section of the Institute of Transportation Engineers. He has served on National Cooperative Highway Research Program panels on highway capacity and is a member of the AASHTO Standing Committee on Highways and the TRB Committee on RRR Standards. He is a registered professional engineer in the state of New York.

RICHARD R. STANDER, JR., is a businessman and President of Mansfield Asphalt Paving Company. At Mansfield Asphalt Paving Company, he served as construction crew foreman, civil engineer, and vice president. He received his bachelor's degree and MBA from Ohio State University and his master's degree from MIT. Mr. Stander served as a part-time lecturer in construction management at Ohio State University from 1969 to 1971. He is a member of the TRB Construction Management Committee, the Ohio Contractors Association, and the Association of Asphalt Paving Technologists. He is a registered professional engineer in Ohio.

JAMES I. TAYLOR is Associate Dean of Engineering, University of Notre Dame. He received his bachelor's and master's degrees from the Case Institute of Technology and his Ph.D. from Ohio State University. He was formerly Chairman of the Department of Civil Engineering, University of Notre Dame, and Director of the Bureau of Highway Traffic at Pennsylvania State University. Dr. Taylor has worked as a consultant to a number of organizations, including the TRB, Goodell-Grivas Incorporated, HRB-Singer, Incorporated, and the Institute For Research. He was formerly President of the North Central Indiana Branch of the American Society of Civil Engineers and the Educational Division of the American Road and Transportation Builders Association. Dr. Taylor is a member of the American Society for Engineering Education, Institute of Transportation Engineers, Transportation Research Board, Sigma Xi, and Tau Beta Pi.

E. DEAN TISDALE is Director of the Idaho Department of Transportation. He is a graduate of the University of Idaho with a bachelor of science degree in forestry (1950) and in civil engineering (1955). Since 1953 Mr. Tisdale has served in the following positions at the Idaho Department of Transportation: Planning Engineer, Deputy State Highway Engineer, State Highway Administrator, and Chief of Engineering Services. He was Vice Chairman of the AASHTO Standing Committee on Highway Traffic Safety, and formerly Chairman of the AASHTO Subcommittee for Traffic Engineering and the AASHTO delegation to the National Committee on Uniform Traffic Control Devices. He is also a member of the AASHTO Executive and Policy Committees, AASHTO Reorganization Task Force, Chairman of the Federal Mandate Review II Task Force, and a member of the University of Idaho Engineering Advisory Board. He is currently Chairman of the Idaho Traffic Safety Commission and a member of the National Society of Professional Engineers and the Idaho Society of Professional Engineers.